AF594941

Feature Papers of Water Resources Management, Policy and Governance

Feature Papers of Water Resources Management, Policy and Governance

Editors

Athanasios Loukas
Luis Garrote

MDPI • Basel • Beijing • Wuhan • Barcelona • Belgrade • Manchester • Tokyo • Cluj • Tianjin

Editors
Athanasios Loukas
Department of Rural and Surveying Engineering,
Aristotle University of Thessaloniki,
54124 Thessaloniki, Greece

Luis Garrote
Department of Civil Engineering: Hydraulics, Energy and Environment,
Technical University of Madrid,
Madrid, Spain

Editorial Office
MDPI
St. Alban-Anlage 66
4052 Basel, Switzerland

This is a reprint of articles from the Special Issue published online in the open access journal *Water* (ISSN 2073-4441) (available at: https://www.mdpi.com/journal/water/special_issues/Management_Policy_Governance).

For citation purposes, cite each article independently as indicated on the article page online and as indicated below:

LastName, A.A.; LastName, B.B.; LastName, C.C. Article Title. *Journal Name* **Year**, *Volume Number*, Page Range.

ISBN 978-3-0365-4945-3 (Hbk)
ISBN 978-3-0365-4946-0 (PDF)

Cover image courtesy of Athanasios Loukas.

© 2022 by the authors. Articles in this book are Open Access and distributed under the Creative Commons Attribution (CC BY) license, which allows users to download, copy and build upon published articles, as long as the author and publisher are properly credited, which ensures maximum dissemination and a wider impact of our publications.
The book as a whole is distributed by MDPI under the terms and conditions of the Creative Commons license CC BY-NC-ND.

Contents

About the Editors

Athanasios Loukas

Prof. Dr. Athanasios Loukas is Full Professor of Engineering Hydrology—Water Resources Management and Development in the School of Rural and Surveying Engineering, Aristotle University of Thessaloniki, Greece. He is the Director of the Master Program "Water Resources" (2018-today). He has served as Director of Transportation and Hydraulic Engineering Department (2019–2021). He is a visiting professor at Colorado State University (U.S.A.) and University of Grenoble Alpes. He has received many academic awards in his professional and academic career, such as the Fulbright Research Scholarship.

He is Section Editor-in-Chief of the MDPI *Water* journal (Section of Water Resources Management, Policy, and Governance) and Editor-in-Chief of EWRA *European Water* journal and Editorial Board Member of Water Resources Management and Scientific Review Engineering and Environmental Sciences journals. He is an active reviewer for many international journals in the fields of hydrology, water resources management, GIS and remote sensing applications, environmental modelling and management and others.

His research interests include: deterministic and stochastic hydrological modelling, analysis, modelling, and forecasting of extreme hydrological events (floods and droughts), climate change impacts on the hydrological processes and water resources, water resources management and modelling, and applications of GIS and remote sensing in hydrology and water resources. He has coordinated and participated in many national and European research projects. He is the reviewer and evaluator of National, International and European research proposals. He supervised and still supervises a large number of undergraduate diploma theses, Master theses and Ph.D. dissertations. He has delivered many keynote speeches and invited talks.

Prof. Loukas is the author and co-author of more than 400 referred international journal publications, conference proceedings publications, and technical reports. His published work has received over 2200 citations, as indicated in the Scopus database with h-index = 29 and over 4200 citations, as indicated in the Google Scholar database (h-index = 36, i10-index = 67).

Luis Garrote

Prof. Dr. Luis Garrote is Full Professor of Hydraulic Engineering at Universidad Politécnica de Madrid. His research focus is the application of hydrological and hydraulic models in water resources planning and management, including floods, droughts, environmental constraints and reservoir operation, with special emphasis on dealing with uncertainties, particularly those connected to global change. His professional record in integrated water resources management includes collaborations with national and international administrations and the private sector in water resources planning, design and operation of hydraulic structures, dam operating rules and safety, hydropower and drought and flood risk management.

Preface to "Feature Papers of Water Resources Management, Policy and Governance"

This book includes the papers published in the *Water* Special Issue entitled "Feature Papers of Water Resources Management, Policy and Governance". The book includes seven papers by invited renowned researchers and engineers to cover various contemporary issues of water resource management, governance, and policy. Water resource management, policy, and governance are great global challenges. These global challenges are current and evident due to competition for limited resources, regional disparities in water supply and affluence, growing global water demand, surface water and groundwater depletion and pollution, and climate-change-induced water stress. Addressing these issues was the motivation for setting up this special collection of papers. The book is addressed to researchers, engineers and scientists, working in the field of water resources. The Guest Editors thank all the contributing invited authors, manuscripts reviewers, assistant editors, and the publication team of MDPI for the great support to the realization of the present book, especially, Rachel Lu, Assistant Editor.

Athanasios Loukas and Luis Garrote
Editors

MDPI

Editorial

Feature Papers of Water Resources Management, Policy and Governance

Athanasios Loukas [1,*] and Luis Garrote [2]

1 Department of Rural and Surveying Engineering, Aristotle University of Thessaloniki, 54124 Thessaloniki, Greece
2 Department of Civil Engineering: Hydraulics, Energy and Environment, Universidad Politécnica de Madrid, 28040 Madrid, Spain; l.garrote@upm.es
* Correspondence: agloukas@topo.auth.gr; Tel.: +30-2310-996103

Water resource management aims to environmentally and economically satisfy the water demands of various water uses in a hydrological basin. The main sectors of water use are agriculture, urban, industry, hydroelectric power production, and preservation of the environment and the ecology of ecosystems. These uses of water require different water volumes and water quality standards and are usually very competitive. Most of the time, it is not possible to cover all water needs in a hydrological basin, due to the limited available water resources. Hence, it is imperative to set water use priorities in a way that serves societal and ecological needs. Managing the water resources and operating the water works may, sometimes, lead to confrontations, deliberations and negotiations. Proper policies and governance for integrated sustainable water resource management are essential. Water resource management, policy and governance are great global challenges due to competition for limited resources, regional disparities in water supply and affluence, growing global water demand, surface water and groundwater depletion and pollution, and climate-change-induced water stress.

This WATER Special Issue (S.I.) titled "Feature Papers of Water Resources Management, Policy and Governance" was set up to collect papers by invited reputable researchers and engineers to cover issues of water resource management, governance, and policy, such as: integrated water resource management, management of water resource systems and water availability, monitoring and protection of water resources, national and international water policy, institutional arrangements, and water law, water economics and commercialization of water, water conflict resolution, public participation, and decision making, water resource management, policy and governance in socially and environmentally sensitive areas and regions. The seven (7) papers of this S.I. cover a wide range of research topics related to water resources management, policy and governance. A short description and discussion of this whole set of experiences from the authors' contribution to this S.I. is provided.

Effective water resource management requires assessments of water availability within a framework of complex institutions and infrastructure used to manage extremely variable stream flow shared by numerous, often competing, water users and diverse types of use. Wurbs [1] uses and updates the Water Rights Analysis Package (WRAP) modelling system. WRAP is fundamental to water allocation and planning in the state of Texas in the United States. The WRAP modelling system combines: (1) detailed simulation of water right systems, interstate compacts, international treaties, federal/state/local agreements, and operations of storage and conveyance facilities, (2) simulation of river system hydrology, and (3) statistical frequency and reliability analyses. The continually evolving modelling system has been implemented in Texas by a water management community that includes the state legislature, planning and regulatory agencies, river authorities, water districts, cities, industries, engineering consulting firms, and university researchers. Environmental flow standards have been integrated into the modelling system and comprehensive state-wide water management. The public domain WRAP software and documentation have

Citation: Loukas, A.; Garrote, L. Feature Papers of Water Resources Management, Policy and Governance. *Water* **2022**, *14*, 2191. https://doi.org/10.3390/w14142191

Received: 5 July 2022
Accepted: 6 July 2022
Published: 11 July 2022

Publisher's Note: MDPI stays neutral with regard to jurisdictional claims in published maps and institutional affiliations.

Copyright: © 2022 by the authors. Licensee MDPI, Basel, Switzerland. This article is an open access article distributed under the terms and conditions of the Creative Commons Attribution (CC BY) license (https://creativecommons.org/licenses/by/4.0/).

been generalized for application in other basins of the world. Application of the modelling system in river basins in Texas, U.S.A. and in other basins of the U.S.A. indicates that the modelling system could be applicable worldwide, and it contributes to integration of water allocation, planning, system operations and research.

Securing water resources for the future is a key issue of global change. This issue is strongly connected with the global population growth, climate change, hydrological cycle, economy, energy production, land use change and pollution generation. Simonovic and Breach [2] present and apply the ANEMI3 model. The ANEMI3 model is an integrated global change assessment model that emphasizes the role of water resources. The paper is focused on the development of global water supplies necessary to keep pace with a growing population and the global economy. A series of experiments have been conducted using the ANEMI3 model, in order to assess: (i) the current role of water supply in the global Earth system; (ii) the level of water stress that can be expected in the future; and (iii) what are the potential effects of water quality on global surface water supply and the distribution of water supply types. The results of model simulations show that surface water resources were sufficient to meet the water demand and water quality has not shown to be a significant factor for the development of surface water supplies. However, these impacts are averaged to a global aggregated scale, and they are likely understated.

Environmental flows are necessary and essential for the preservation of river and riparian ecosystems below dams and hydropower projects. However, maintaining environmental flows has faced considerable resistance and caused conflicts among different stakeholders. Appropriate solutions should be examined. Ruan and associates [3] present a study and analysis of questionnaires and interviews to determine the key conflicts in the implementation of environmental flows in a small-scale hydropower project in China and to propose potential solutions. Three factors have been selected as the main reasons for conflicts, namely, economics, stakeholders' skepticism, and technology according to the international literature. The study uses online questionnaires and interviews with owners of small-scale hydropower projects, government administrators, and the public in Fujian Province, China. The results showed that the main hindrance for the implementation of environmental flows was the potential economic loss resulting from reductions in electricity production, stakeholder's' skepticism, technical difficulties, and a lack of the government supervision. Diversion-type projects pose the largest losses of electricity production after the release of environmental flows, and by adopting a 10% of mean annual flow as minimum target, most small-scale hydropower projects obtain low marginal profits without compensation. The authors proposed an appropriate payment for ecosystem services by introducing an economic compensation program for different types of small-scale hydropower projects scaled by potential losses in electricity generation. Under such a scheme, the government, hydropower project owners, and electricity consumers share the cost of economic losses from a reduction in electricity production. The paper also presents recommendations for policymakers, officials, and researchers for conflict mitigation when implementing environmental flows.

The understanding and realization of the complexity of water governance beyond an empirical concept is significant. Gumeta-Gómez and associates [4] propose a Water Governance Complexity Framework to address the complexity of water governance. Through a literature review, rapid surveys, and 79 semi-structured interviews, the authors propose how this framework may become operational using different representations. The framework has been applied to the urban water supply system of Oaxaca, Mexico. The authors found legal pluralism and diverse formal and informal stakeholders in a multilevel structure in rural communities of Oaxaca, where the state plays a partially absent role in the water supply. Four modes of governance at the local level were identified, resulting from seven trajectories of institutional change. These trajectories result from linear (alignment) and nonlinear (resistance and adaptation) interactions between local, state, and national institutions over different periods. The authors provide a pragmatic framework to understand complexity through the organization and historical configurations of water governance that may be applied globally, providing a necessary starting point and solid foundation for the creation of new water policies and law reforms or transitions to the polycentric governance model to ensure the human right to water and sanitation.

The preservation of water ecosystems is imperative in the framework of water resource management. Investigations about changes in ecosystems and their relevant water environments under rapid changes in land use can provide valuable information to formulate sustainable protection and development strategies. Zhang and associates [5] present a study on the preservation of the ecosystem of the mulberry-dyke-fish ponds, which are a representative traditional eco-agriculture in the Greater Bay Area of Guangdong–Hong Kong–Macao (GBA). The study combines supervised classification and visual interpretation approaches using Landsat images obtained after 1986. A water intensity index and a synthesized index are used to identify spatial patterns of changes in the ponds in the GBA over the past 40 years. The results indicate that during the period 1986–2013, the total surface area of the ponds in the GBA increased significantly and reached its peak in 2013 with a total increase of 84.63%. After this period, the total surface area of the pond showed a downward trend with a total decrease of approximately 31.34%. The year 2013 was identified as the critical year of the changes. It seems that human activities have continuously influenced the spatial distribution and size of fishponds in the past 40 years. The fishponds had transformed from near-natural ponds with different sizes and a near-natural random distribution in the early stage into an artificial distribution and an artificial shape. Land use changes, industrial transfer, government guidance and financial motives have been identified as the major drivers for the changes. This shrinking trend in the ponds will continue in the future, if no effective measures are taken.

Natural hazards have caused significant damages to natural and manmade environments during the last few decades. Hydro-meteorological hazards are among the most destructive hazards and are considered responsible for the loss of human lives, infrastructure damages and economic losses [6]. Droughts affected 52.7 million people worldwide in year 2021 and 67.5 million people worldwide, on average per year, for the period 2001–2020 [7]. Drought is one of the most damaging natural hazards on the Iberian Peninsula, causing significant socioeconomic and environmental problems. Five (5) major river basins are transboundary river basins between Portugal and Spain. Cooperation between the two countries is needed to prevent the adverse impacts of droughts. However, in terms of drought planning and management the two countries are clearly in different stages. Portugal approved a national drought plan in 2017, while Spain has already had drought plans in place for all River Basin Districts since 2007 and approved an updated version of these plans in 2018. The Spanish drought plans currently in place foresee two sets of indicators: prolonged drought and water scarcity indicators. Maia and associates [8] present the definition of similar indicators for the Portuguese part of the Minho and Lima transboundary river basins, according to European guidelines and in common with Spain, with the aim of developing a joint international drought management plan for these basins. For the period from October 1980 to September 2017, the comparison of the indicators obtained for the Portuguese parts of the basins with those obtained for the corresponding Spanish parts shows a similarity in the occurrence of drought and water scarcity in both parts of the basins, although with a higher prevalence of water scarcity situations in the Spanish part of the Lima river basin. The work presented in the paper has been developed in close collaboration with the competent authorities of the river basin districts of both countries, with the aim to be a prototype for the definition of new and comparable indicators of drought and operational scarcity. Therefore, this work is a starting point for the creation of common tools for integrated management of drought in transboundary basins in the Iberian Peninsula.

Sustainable water resources management implies the study of all interrelated parameters (e.g., social, environmental, economic, engineering and political) in a comprehensive way. Although Greece is listed in the international rankings as a water-rich country, it has significant water problems due to its high temporal and spatial variation in the distribution of water resources and its unsustainable management practices, characterized by a fragmented and sector-oriented water management system. This problem has been significantly improved by the adoption of the EU Water Framework Directive (EU WFD) and the development of management plans at the river basin scale [9]. However, because of the effects of climate change, there is still a long way to go, and substantial changes are needed in order to reach sustainability. In this sense, adaptation is a vital response toward sustainability. Kolokytha [10] presents an analysis of water resources management and the application of EU WFD in the Mediterranean region and Greece. The paper focuses on the example of the Mygdonia basin located in Northern Greece. The agricultural basin of Mygdonia is a case study of a highly negative water balance system that highlights the shortcomings of both water management and adaptation in Greece. Analysis of the hydrology of the basin, as well as the climate projections until 2100, revealed the urgent need for concerted actions. A set of different adaptation strategies for development was applied and assessed for their effectiveness. According to the results of this research, integrated watershed management is a prerequisite for a successful adaptation policy. Radical reform is needed in the agricultural sector by decreasing agricultural land and changing the crop pattern. The study concludes that managing water demand is the solution rather than the development of water supply projects.

In conclusion, this Special Issue contains seven (7) invited papers with important results, covering several aspects of water resources management, policy and governance. Increasing demand for water, under the pressure of climate change impacts, is forcing water scientists and engineers to improve and develop new methods and approaches for integrating water resources management and protection, develop appropriate policies and define feasible governance structures. The results offer insights for further multi-methodological, multi-disciplinary and multi-purpose research. There are still challenges to be accepted and overcome to ensure a sustained and sufficient supply of good quality water for future generations.

Author Contributions: A.L. and L.G. have conceived the structure and wrote the manuscript. All authors have read and agreed to the published version of the manuscript.

Funding: This research received no external funding.

Acknowledgments: In the Guest Editors thank all the authors of the published papers and the manuscripts reviewers.

Conflicts of Interest: The authors declare no conflict of interest.

References

1. Wurbs, R.A. Institutional Framework for Modeling Water Availability and Allocation. *Water* **2020**, *12*, 2767. [CrossRef]
2. Simonovic, S.P.; Breach, P.A. The Role of Water Supply Development in the Earth System. *Water* **2020**, *12*, 3349. [CrossRef]
3. Ruan, Q.; Wang, F.; Cao, W. Conflicts in Implementing Environmental Flows for Small-Scale Hydropower Projects and Their Potential Solutions—A Case from Fujian Province, China. *Water* **2021**, *13*, 2461. [CrossRef]
4. Gumeta-Gómez, F.; Sáenz-Arroyo, A.; Hinojosa-Arango, G.; Monzón-Alvarado, C.; Mesa-Jurado, M.A.; Molina-Rosales, D. Understanding the Complexity of Water Supply System Governance: A Proposal for a Methodological Framework. *Water* **2021**, *13*, 2870. [CrossRef]
5. Zhang, W.; Cheng, Z.; Qiu, J.; Park, E.; Ran, L.; Xie, X.; Yang, X. Spatiotemporal Changes in Mulberry-Dyke-Fish Ponds in the Guangdong-Hong Kong-Macao Greater Bay Area over the Past 40 Years. *Water* **2021**, *13*, 2953. [CrossRef]
6. Tsakiris, G. Flood risk assessment: Concepts, modelling, applications. *Nat. Hazards Earth Syst. Sci.* **2014**, *14*, 1361–1369. [CrossRef]
7. Centre for Research on the Epidemiology of Disasters. Disasters (CRED) in Numbers 2021. CRED 2022. Available online: https://cred.be/sites/default/files/2021_EMDAT_report.pdf (accessed on 15 June 2022).
8. Maia, R.; Costa, M.; Mendes, J. Improving Transboundary Drought and Scarcity Management in the Iberian Peninsula through the Definition of Common Indicators: The Case of the Minho-Lima River Basin District. *Water* **2022**, *14*, 425. [CrossRef]
9. Implementation of River Basin Management Plans—Greece—Environment—European Commission. Available online: https://ec.europa.eu/environment/water/participation/map_mc/countries/greece_en.htm (accessed on 8 December 2021).
10. Kolokytha, E. Adaptation: A Vital Priority for Sustainable Water Resources Management. *Water* **2022**, *14*, 531. [CrossRef]

Article

Institutional Framework for Modeling Water Availability and Allocation

Ralph A. Wurbs

Department of Civil and Environmental Engineering, Texas A&M University, College Station, TX 77843, USA; r-wurbs@tamu.edu; Tel.: +1-979-845-3079

Received: 3 September 2020; Accepted: 1 October 2020; Published: 4 October 2020

Abstract: Effective water resources management requires assessments of water availability within a framework of complex institutions and infrastructure employed to manage extremely variable stream flow shared by numerous, often competing, water users and diverse types of use. The Water Rights Analysis Package (WRAP) modeling system is fundamental to water allocation and planning in the state of Texas in the United States. Integration of environmental flow standards into both the modeling system and comprehensive statewide water management is a high priority for continuing research and development. The public domain WRAP software and documentation are generalized for application any place in the world. Lessons learned in developing and implementing the modeling system in Texas are relevant worldwide. The modeling system combines: (1) detailed simulation of water right systems, interstate compacts, international treaties, federal/state/local agreements, and operations of storage and conveyance facilities, (2) simulation of river system hydrology, and (3) statistical frequency and reliability analyses. The continually evolving modeling system has been implemented in Texas by a water management community that includes the state legislature, planning and regulatory agencies, river authorities, water districts, cities, industries, engineering consulting firms, and university researchers. The shared modeling system contributes significantly to integration of water allocation, planning, system operations, and research.

Keywords: water allocation; planning; river/reservoir systems; water availability modeling

---

1. Introduction

Effective water allocation and management requires an understanding of the reliabilities at which various quantities of water can be provided under various conditions. Modeling and analysis strategies for quantifying capabilities for supplying water needs are explored in this paper based on the experience of the Texas water management community in developing and applying a legislatively mandated water availability modeling system to support statewide planning and water allocation. The modeling system has been expanded and improved continually over the past twenty years to address evolving water management strategies and issues. Current research, development, and implementation priorities include incorporation of legislatively mandated environmental flow standards in both the modeling system and actual water management. The Brazos River Basin represents the inaugural application of the latest version of the modeling system with expanded features added to incorporate environmental flow standards and serves as a case study to illustrate the concepts and issues discussed in this paper.

The river/reservoir system simulation and frequency/reliability analysis methods presented in this paper are implemented in a comprehensive, flexible modeling system developed at Texas A&M University (TAMU) called the Water Rights Analysis Package (WRAP) [1–6]. The public domain software package is generalized for application anywhere in the world and has been employed in various other countries and states but not to the same extent as its application in Texas. A water availability modeling (WAM) system maintained by the Texas Commission on Environmental Quality (TCEQ) consists of WRAP and input datasets for all of the river basins of Texas [7,8].

Generalized computer modeling systems have played increasingly important roles in various aspects of water resources planning and management throughout the world over the past several decades [9,10]. The term "generalized" is used here to mean that the software is designed to be applied to real-world systems of various configurations at different locations by professional practitioners other than the original model developers. Generalized models should be thoroughly tested, clearly documented, and conveniently accessible. Wurbs [11], Lababie [12], Rani and Moreira [13], Lund et al. [14], and many others provide reviews of the massive literature on modeling multiple-purpose river/reservoir system operations. Most of the numerous river basin management models reported in the literature are not generalized.

Wurbs [15,16] reviews the literature on modeling reservoir/river system management and compares WRAP with other generalized modeling systems, focusing specifically on HEC-ResSim [17], RiverWare [18], and MODSIM [19]. RiverWare is marketed by the Center for Advanced Decision Support for Water and Environmental Systems (CADSWES) for a licensing fee. CADWES also provides consulting services to support application of RiverWare. HEC-ResSim, MODSIM, and WRAP software and documentation can be downloaded free-of-charge from their websites. HEC-ResSim, developed at the U.S. Army Corps of Engineers (USACE) Hydrologic Engineering Center (HEC), is applied nationwide to support operations of USACE multiple-purpose reservoir system operations, particularly flood control operations. MODSIM, developed at Colorado State University, is based on linear programming and has been applied to river/reservoir systems in many countries including systems operated by the U.S. Bureau of Reclamation in the United States. WRAP provides particular flexibility for modeling prior appropriation water rights permit systems and other institutional water allocation mechanisms. WRAP is designed for efficient modeling and analysis of large complex river systems with many hundreds of reservoirs and water users [15,16].

Expanded capabilities for assessing water availability and supply reliability have been essential to recent improvements in water management in Texas. Strategies and methods employed in Texas are applicable worldwide. Various issues that are still not fully resolved in Texas are also important in other regions of the world. The objective of this paper is to employ the Texas experience to outline water availability and allocation assessment practices proven to be effective and to highlight key complexities that have been successfully addressed along with needs for further advances. Computer-based modeling and analysis are integrated with water allocation and management.

2. Water Resources Planning, Allocation, and Management in Texas

The geographic, climatic, hydrologic, and economic diversity that spans the state of Texas combined with high population growth and progressive water management practices makes Texas an excellent laboratory for investigating water management strategies and assessment tools that are generally applicable throughout the United States and the world. Motivated by continually intensifying demands on limited water resources, the state has implemented an array of strategies over the past twenty years that have greatly improved water management [8,20,21]. Greatly expanded water availability modeling capabilities have provided essential decision support.

The 682,000 km^2 area of Texas (Figure 1) is comprised of 15 major river basins and eight coastal basins located between the major rivers. Mean annual precipitation increases from west to east across Texas from 20 to 145 cm. The population increased from 3,060,000 people in 1900 to 20,950,000 in 2000, to 25,390,000 in 2010 and 29,700,000 in 2020, and is projected by the Texas Water Development Board (TWDB) to increase to 46,360,000 by 2060 [20]. Declining groundwater supplies combined with population growth are resulting in intensified demands on surface water resources [8,20].

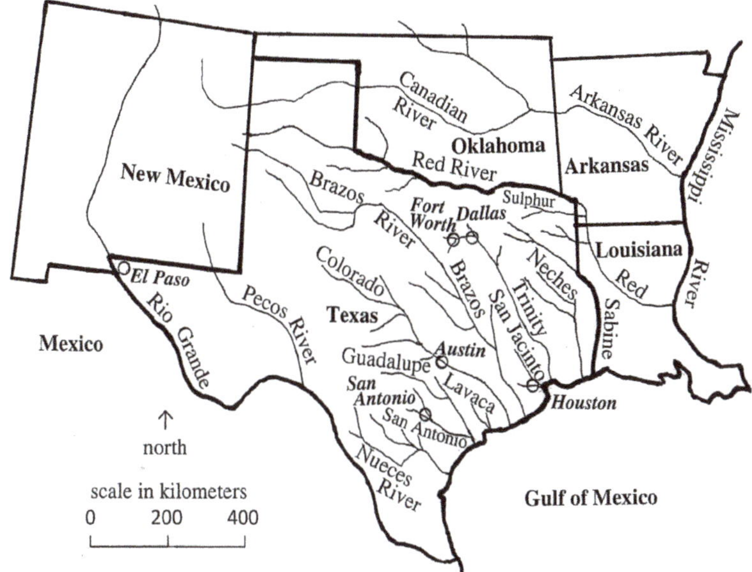

Figure 1. Map of major rivers and largest cities in Texas.

Ground and surface water each currently provide about half of the total water supply in Texas, with a shift toward less groundwater. Groundwater is used throughout the state, though agricultural irrigation supplied from the Ogallala Aquifer in northwest Texas accounts for the largest portion of the groundwater use. Groundwater rights in Texas have been based on the common law rule allowing landowners to pump unlimited quantities of water from under their land [8,21,22]. Most land in Texas is privately owned. Increased regulation of groundwater is evolving over time primarily through the establishment of local groundwater conservation districts. The 102 diverse groundwater districts established to date cover all or part of 184 of the 254 counties of the state. These districts encourage water conservation, protect water quality, and to a limited but growing extent regulate pumping.

This paper focuses on water in streams and reservoirs. Surface water is owned by the state. A state agency, the TCEQ, regulates the diverse use of surface water by numerous users.

Allocation of stream flow in Texas evolved over several centuries of rule by Spain, Mexico, the Republic of Texas, and the State of Texas into an unmanageable assortment of diverse water rights based on various versions of the riparian and prior appropriation doctrines [22]. The waters of the Rio Grande are allocated between the U.S. and Mexico by a 1944 treaty. The economy of the Lower Rio Grande Valley is based on irrigated agriculture. A severe drought during 1950–1957 motivated massive lawsuits that resulted in judicial allocation of rights to use the Texas share of the Rio Grande. The Water Rights Adjudication Act of 1967 initiated a 25-year process of consolidating the numerous water rights for the remainder of Texas into a permit system. Texas participates with neighboring states shown in Figure 1 in interstate compacts for the following rivers and effective dates: Rio Grande—1939, Pecos—1948, Canadian—1952, Sabine—1954, and Red—1980. All of these surface water allocation schemes are reflected in the water rights system and simulated in the water availability model (WAM) system maintained by the Texas Commission on Environmental Quality (TCEQ).

Surface water rights are granted by a state license, or permit, which allows the holder to divert a specified amount of water annually at a specific location, for a specific purpose, and to store water in reservoirs of specified capacity. Any organization or person may submit an application to the TCEQ for a new water right or to change an existing water right at any time. The TCEQ will approve the permit application if unappropriated water is available, existing water rights are not impaired, efficient water conservation will be practiced, and proposed actions are consistent with regional water plans. A permit holder does not own surface water but only a right to use the water. However, water rights can be sold, leased, or transferred. Such transfers are encouraged but require TCEQ permit approval.

Water management occurs within an institutional setting that includes laws enacted by the Texas Legislature that are implemented collaboratively by government agencies, private industry, stakeholders, consulting engineering firms, university researchers, and the general public. Several legislatively mandated programs have motivated or necessitated advances in water availability modeling capabilities to support water planning, development, allocation, and management.

Omnibus water management legislation enacted by the Texas Legislature as its 1997 Senate Bill 1 (SB1) authorized a statewide and regional water planning process and creation of the WAM system to support planning and water allocation [7]. The Texas Water Development Board (TWDB) has been conducting statewide planning since the 1950s. The 1997 SB1 created a structured planning strategy that emphasizes local and regional participation. Sixteen regional water plans developed by planning groups supported by the TWDB and consulting firms and a consolidated statewide plan developed by TWDB staff in collaboration with the water management community are updated in a five-year planning cycle with a 50-year future planning horizon [20]. Reports documenting the 2002, 2007, 2012, and 2017 water plans are available at the TWDB website [23]. Work on the updated 2022 regional and statewide plans is progressing.

The 2001 Senate Bill 2 created the Texas Instream Flow Program to advance the science of environmental flows and associated management strategies [24]. The 2007 Senate Bill 3 (SB3) created a process for establishing environmental flow standards (EFS) based on best currently available science and incorporating these standards in the WAM System [25]. Periodic updates to flow standards are anticipated with advances in instream flow science and management strategies. Integration of SB3 environmental flow standards (EFS) in water management and water availability modeling is a major focus of continuing efforts to expand WRAP and the Texas WAM system.

The flow of rivers in Texas, like other regions throughout the world, is characterized by great variability that includes the extremes of intense floods and severe multiple-year droughts combined with seasonal and continuous fluctuations [26]. Large reservoir storage capacities are essential for managing flow variability and uncertainties regarding future water availability. Numerous water users share limited stream flow and reservoir storage that is used for a diversity of purposes. Multiple-purpose, multiple-reservoir system operations are fundamental to effective water management. Preserving the vitality of riverine ecosystems while supplying water, electrical energy, and other needs of growing populations and economies is a global challenge [27–30] as well as a legislatively mandated requirement in Texas [25].

3. Water Rights Analysis Package (WRAP) and Water Availability Modeling (WAM) System

The monthly version of the WRAP modeling system is routinely applied in Texas with simulation input datasets from the WAM system maintained by the TCEQ. The generalized WRAP combined with a simulation input dataset for a particular river basin is called a water availability model (WAM). Model users modify the Texas WAM system datasets to reflect water use requirements, proposed projects, and management strategies of interest. For applications outside of Texas, model users develop their own input datasets for river/reservoir systems of interest. Input datasets range from small and simple to extremely large and complex. The monthly WRAP has been routinely applied for many years while continually being expanded and improved. Integration of SB3 EFS into the WAMs and comprehensive water management has motivated development of daily modeling capabilities that are now transitioning from research and development to implementation.

3.1. Evolution of the WRAP Modeling System

Software, manuals, datasets for examples in the manuals, and other information are available free-of-charge at the TAMU WRAP website [31], which links with the TCEQ WAM website.

The manuals [1–6] are published as technical reports by the Texas Water Resources Institute (TWRI) of the Texas A&M University (TAMU) System. Other WRAP-related technical reports are also available at the TWRI website [32]. The reference manual [1] includes a Bibliography of WRAP-Related Publications that lists 18 M.S. theses and ten Ph.D. dissertations by TAMU graduate students and many reports and journal and conference papers.

The predecessor to WRAP, called TAMUWRAP, was developed in a project funded by a federal/state cooperative research program administered by the U.S. Department of Interior and TWRI with the Brazos River Authority (BRA) serving as a nonfederal sponsor [1,33]. The modeling system has been continually improved and expanded since its implementation in the TCEQ WAM System [7]. The TCEQ has sponsored WRAP research and development at TAMU continuously during 1997–2003 and 2005–2021, concurrently with other WRAP-related research projects funded by other agencies. Development of methods incorporated in WRAP and research studies at TAMU using WRAP to explore various water management issues have been funded by the TCEQ, TWDB, TWRI, BRA, Texas Advanced Technology Program, U.S. Army Corps of Engineers, U.S. Department of Energy, National Institute for Environmental Global Change, and other agencies [1].

The components of WRAP routinely applied with Texas WAM datasets are based on a monthly computational time step. The May 2019 WRAP software and manuals accessible at the WRAP website expand the monthly modeling system to also include daily modeling capabilities with monthly-to-daily naturalized flow disaggregation, flow routing, forecasting, flood control reservoir operations, and instream flow standards with subsistence, base, and high-pulse flow components.

A driving motivation for the daily modeling system is the 2007 Senate Bill 3 (SB3) requirement that environmental flow standards (EFS) be established and incorporated in the TCEQ WAM system [25]. As of late 2020, SB3 EFS have been incorporated in developmental daily versions of the Brazos, Trinity, and Neches WAMs to compute daily instream flow targets that are summed to monthly targets for incorporation in the WRAP input dataset for the monthly models [34–36]. These daily WAM datasets and detailed technical reports are available at the TAMU WRAP website [31].

3.2. Texas Water Availability Modeling (WAM) System

The WAM System was created pursuant to the 1997 SB1 by the TCEQ, TWDB, other partner agencies, and contractors consisting of consulting engineering firms and university researchers [7]. Authorized use and current use scenario versions of 20 WRAP simulation input datasets covering all Texas river basins, an array of other information, and a link to the TAMU WRAP website are accessible at the TCEQ WAM website [37].

The TCEQ is the lead agency in maintaining the WAM System along with administrating the water rights permit system and interstate river basin compacts. Water right permit applicants, or their consultants, are required by the TCEQ to apply the WAMs to assess water supply reliabilities of proposed actions and the impacts on the reliabilities of all other water users. TCEQ staff apply the modeling system in evaluating permit applications. The TCEQ usually has over 200 water right permit applications under review at any time. Many are proposed modifications to existing permits.

The TWDB and 16 regional planning groups apply the WAMs in the regional and statewide planning process established by the 1997 SB1. River authorities and other entities apply the WAMs in operational planning studies and other endeavors. The modeling system has also been applied in U.S. Army Corps of Engineers (USACE) regulatory activities, environmental flow studies, project feasibility studies, university research studies, and other water management endeavors.

The 15 major river basins and eight coastal basins of Texas are modeled as 20 WAMs, with three WAMs containing two adjoining basins. Activities of numerous water management entities operating over 3400 dams/reservoirs and other constructed facilities in accordance with treaties between the U.S. and Mexico, five interstate compacts, two water right permit systems with 6200 active permits, federal water supply contracts, and other institutional arrangements are simulated.

Authorized and current use scenario datasets are available at the TCEQ WAM website for each of the 20 WAMs. The authorized use scenario is based on the premise that all water right permit holders use the full amounts to which they are legally entitled, subject to water availability. Many permits include projected future water needs. The current use scenario represents actual recent water use. The TWDB has developed WAM datasets representing projections of future water needs.

The modeling system contributes greatly to water management and continues to be expanded to address various issues. Modeling support for establishing SB3 EFS is currently a priority research, development, and implementation focus, along with improving capabilities for water management during drought and more efficiently updating simulation input datasets.

4. Modeling and Analysis Methodologies

WRAP simulates capabilities of river/reservoir systems in meeting specified water management, regulation, and use requirements for given sequences of naturalized stream flows and reservoir net evaporation less precipitation rates. A specified scenario of water management is combined with natural historical hydrology. Since the future is unknown, historical hydrology is used to statistically capture the hydrologic characteristics of a river basin. The water management and use scenario might be actual current water use, projected future conditions, the premise that all permit holders use their full authorized amounts, or some other scenario of interest. Simulation results are organized in optional formats including tabulations and plots of entire time sequences, summary tables, water budgets, frequency relationships, and various types of reliability indices. Water management capabilities are expressed in terms of the likelihood (reliability) of meeting water supply targets or portions thereof and stream flow and reservoir storage frequency relationships.

The WRAP modeling system includes executable computer programs that perform the functions outlined as follows.

1. WinWRAP is a user interface for managing programs and data files within Microsoft Windows.
2. Development of Hydrology Input Data for the Simulation Model:
    - Program HYD described by the Hydrology Manual [4] develops and updates SIM input files of monthly naturalized stream flows and reservoir net evaporation-precipitation rates.
    - Program DAY documented by the Daily Manual [5] is used to calibrate routing parameters and otherwise compile daily hydrology input data for SIMD.
    - The Hydrologic Engineering Center (HEC) Data Storage System Visual Utility Engine (DSS-Vue) [38] is used to compile, analyze, and manage times series datasets.
3. Simulation of the River/Reservoir Water Management/Allocation/Use System:
    - Program SIM performs monthly simulations as described by the Reference, Users, and Fundamentals Manuals [1–3].
    - Program SIMD performs daily simulations as described in the Reference, Users, and Daily Manuals [1,2,5].
4. Tracking Salinity through the River/Reservoir System:
    - Program SALT performs a salinity simulation by combining the results of a SIM simulation with a salinity input file [6,39].
5. Post-Simulation Analyses of Simulation Results:
    - Program TABLES reads SIM, SIMD, and SALT simulation input and results, performs frequency and reliability analyses, and creates a variety of tables to organize, summarize, analyze, and display simulation results [1–3].

- HEC-DSSVue [38] reads HYD, SIM, SIMD, TABLES, and SALT DSS input and output files containing time series of hydrology input or simulation results, prepares plots, and performs mathematical and statistical analyses and other data management functions.

The well-established but still evolving WRAP simulation model SIM performs water accounting computations using a monthly time step. SIMD is a recently developed expanded version of SIM that performs the simulation computations using a daily time step. The daily SIMD maintains all capabilities of the monthly SIM while incorporating additional features for monthly-to-daily disaggregation of stream flows and water use targets, flow routing, forecasting, flood control reservoir operations, and tracking high-pulse flows defined by environmental flow standards.

The USACE Hydrologic Engineering Center (HEC) Data Storage System (DSS) has been fully integrated in WRAP for managing time series data. The latest versions of the WRAP programs create, read, and store data in DSS files. The DSS interface HEC-DSSVue [38] is an integral component of WRAP. The HEC of the USACE developed and maintains several generalized modeling systems that are extensively used by government agencies, engineering firms, and universities throughout the United States and abroad. HEC-DSS and its HEC-DSSVue interface are shared by HEC models and have also been incorporated in other non-HEC modeling systems, including WRAP.

4.1. SIM and SIMD Simulation Models

The spatial configuration of a river system is defined in the simulation model by a set of control points, with the next downstream control point being specified for each control point. All reservoirs, water supply diversions, return flows from surface and groundwater supply sources, hydroelectric power plants, instream flow requirements, and other system components are assigned control point locations. Essentially, any configuration of stream tributaries and conveyance systems may be modeled. The 20 WAMs contain over 12,000 control points of which about 500 are primary. The term "primary" control point refers to a site, usually a stream flow gauge, at which naturalized stream flows are stored in the WAM input datasets. Naturalized flows at primary control points are developed by adjusting observed flows to remove the effects of human water development and use. Naturalized flows at all other control points are computed in the simulation based on the naturalized flows at the primary control points and watershed parameters contained in the WAM datasets.

Regulated and unappropriated flows are computed in the simulation for all control points. Regulated flows represent the stream flows hypothetically occurring when historical naturalized flow sequences are repeated with the water use scenario reflected in the WAM. Unappropriated flows are the stream flows still remaining after all water rights in the WAM are allocated their appropriate shares to supply their storage and use targets. Unappropriated flows may be less than regulated flows due to instream flow requirements and appropriations by senior water rights at downstream sites.

The term "water right" is used in WRAP to refer to a set of water use requirements and associated constructed facilities and operating rules designed to supply the water use requirements. Many water right permits are modeled simply as WRAP water rights. However, a complicated actual water right permit may be simulated with multiple "model water rights". Water use requirements and facilities that are not associated with water right permits are also modeled as "model water rights". Flexibility is provided for simulating complicated water supply, hydropower, and instream flow target-setting criteria and reservoir system operating rules.

Texas, like most states in the western half of the United States, has a water rights system based on the prior appropriation doctrine [21,22]. Priorities are based on dates specified in the 6200 permits reflecting when the right was initially established. Most of the water rights in the WAMs reflect this priority system. However, the generalized WRAP simulation model includes flexible capabilities that include various options for assigning priorities. Subordination agreements that circumvent water right priorities are modeled. One WRAP option assigns priorities in upstream-to-downstream sequencing, modeling the riparian doctrine common in the eastern half of the U.S.

The monthly SIM and daily SIMD simulation computations are performed in a water rights priority sequence that is embedded within a computational time step loop. SIM/SIMD execution begins with reading and organizing input data. Water rights are sorted into priority order based on priority numbers and/or other user-defined options. Naturalized flows provided as input at primary control points are distributed to all other sites within the simulation based on watershed parameters. For each sequential month or day, water accounting computations are performed as each set of water use requirements (water right) is considered in priority order. Water allocation and management are modeled by accounting procedures within the water rights priority loop.

SIM or SIMD simulation results include time series of any of the computed variables. SIM generates only monthly quantities, while SIMD produces daily quantities and monthly summations of the daily quantities. The model-user selects the control points, water rights, and reservoirs for which simulation results are recorded. The simulation results time series variables include: naturalized, regulated, and unappropriated flows, stream flow depletions, and return flows for each selected control point; channel losses and channel loss credits for each selected control point representing the reach below the control point; storage volume, surface elevation, net evaporation, inflows, releases, diversions, and hydroelectric energy at each reservoir; diversion targets and shortages, return flows, available stream flows, stream flow depletions, and storage for each selected water supply right; hydropower targets, firm energy produced, secondary energy produced, energy shortages, and storage for each hydroelectric right, and flow target and shortage for each instream flow right.

The simulation model can be executed in either conventional long-term analysis or short-term conditional reliability modeling (CRM) modes. In the long-term simulation mode normally employed, a specified water management/use scenario is combined with naturalized flows and net reservoir evaporation rates covering the entire hydrologic period-of-analysis in a single simulation. The results are used to generate water supply reliability and stream flow and reservoir storage metrics without reference to present storage contents. In the short-term CRM mode, the hydrologic input is divided into multiple sequences. The simulation is automatically repeated with each hydrologic sequence starting with the same specified initial storage condition. Tables of frequency and reliability metrics from the simulation results are computed with program TABLES. For example, in a CRM analysis, the estimated probabilities of reservoir storage contents reaching various levels any specified number of months in the future conditioned upon specified initial storage levels can be computed [1,2,40].

The simulation model also has options that involve automated repetitions of the complete long-term simulation. A dual simulation option is useful in modeling multiple rights with different priorities associated with the same reservoir system. Another option sets reservoir storage contents at the beginning of a second simulation equal to the storage at the end of an initial simulation.

The TCEQ WAM System is appropriately and effectively constructed based on a monthly computational time step, which is generally optimal for most WAM applications. However, daily computations are needed to model reservoir operations during floods and to incorporate SB3 environmental flow standards (EFS), particularly high-flow pulse components, in the WAMs. The primary differences between daily SIMD and monthly SIM simulation models are as follows.

Flow rates that vary continuously over time in the real world are modeled as volumes occurring during discrete time intervals. Variability is reduced with a larger flow rate averaging time interval. Maximum flood peaks are lowered and minimum flows during low flow periods increase. Monthly flows are less variable than daily flows. Reliabilities of rights with large reservoir storage capacities are less sensitive to time step. Differences are more pronounced for rights with minimal or no storage.

Outflow equals inflow with no attenuation in a monthly SIM simulation whenever a reservoir conservation (water supply and hydropower) pool is full. SIMD simulates flood control operations of any number of reservoirs based on allowable flows at any number of downstream control points. High-flow pulses are also tracked in daily modeling of environmental flow standards.

SIMD disaggregates monthly naturalized flows based on patterns defined by inputted daily flow hydrographs while maintaining the original monthly volumes. Water supply diversions, return flows, reservoir releases, and storage refilling result in changes in stream flows at downstream locations. Flow changes propagate through the stream system in the same month in SIM. Routing in SIMD refers to the downstream propagation of these changes to stream flow. A lag and attenuation routing method is employed in SIMD. A reverse routing algorithm is also applied to replicate the effects of routing in the procedure for forecasting flow availability.

Flow forecasting makes daily computations in SIMD much more complicated than a monthly simulation. Senior water users may be adversely affected by actions of upstream junior users occurring one or more days earlier. Likewise, flood control reservoir operations are based on making no releases that contribute to flows exceeding maximum non-damaging flow limits at downstream gauges that may be located several days of flow travel time below the dam. For each day of the SIMD simulation, the final simulation is preceded by a forecast simulation covering a future forecast period that generates stream flow availability information for that current day.

4.2. Water Availability and Supply Reliability Metrics

The programs TABLES and HEC-DSSVue are used to organize SIM or SIMD simulation results in various user-specified formats, including time series plots or tabulations of selected variables, water budgets, statistical summaries, and various types of frequency relationships and reliability indices.

Options employing either relative counts or probability distribution functions are employed in TABLES and HEC-DSSVue to develop frequency relationships. Relative frequency is expressed by Equation (1) or Equation (2), where m is the rank and N is the sample size. The sample size N is the number of days, months, or years in the period-of-analysis and the rank m is the number of periods during the simulation that a particular flow, storage, or other quantity is equaled or exceeded.

$$Exceedance\ Frequency = \frac{m}{N}\,(100\%) \tag{1}$$

$$Exceedance\ Frequency = \frac{m}{N+1}\,(100\%) \tag{2}$$

Frequency analyses can be performed with WRAP for any time series variable, including any of the numerous simulation input and simulation results variables, variables derived therefrom, or other variables. Equation (1) is commonly applied with stream flow and reservoir storage quantities. With a 1940–2017 period-of-analysis, N is 936 for monthly or 28,490 for daily series of flow or storage quantities and 78 for annual series of July (or any specific month) flow or storage volume. Frequency formula, options Equations (1) and (2) are usually applied for the typically large values of N in WRAP analyses. The log-normal or log-Pearson type III probability distribution options are often applied with annual series generated in a daily SIMD simulation study, such as the minimum or maximum daily stream flow or reservoir storage volume in each year or the minimum or maximum 7-day, 30-day, or any other period stream flow in each year.

The terms "target", "demand", "need", and "requirement" are used interchangeably and may refer to either water supply for municipal, industrial, agricultural, or other types of water use or hydroelectric energy generation. Volume and period reliabilities provide concise metrics for measuring capabilities for meeting water supply diversion and hydroelectric energy generation requirements. Volume reliability (R_V) is the ratio of volume of water supplied or energy produced (v) to the target (V), converted to a percentage, Equation (3). Period reliability is the percentage of the total number of periods of the simulation during which the specified target is either fully supplied or at least a specified percentage of the target is supplied. Period reliability (R_P) is computed by TABLES from the results of a SIM or SIMD simulation, such as Equation (4), where n denotes the number of periods (days, months, years) during the simulation for which a specified percentage of the demand target is met, and N is the total number of periods considered.

$$R_V = \frac{v}{V}(100\%) \tag{3}$$

$$R_P = \frac{n}{N}(100\%) \tag{4}$$

R_P is an expression of the percentage of time that the full demand target or a specified percentage of the demand target can be supplied. Equivalently, R_P represents the likelihood or probability of the target being met in any randomly selected month or year. Reliabilities may be tabulated with the WRAP program TABLES for all or selected individual water rights, the aggregation of all rights associated with individual control points or reservoirs, or user-selected groups of water rights.

A shortage volume in a particular month is the water supply diversion target less the simulated actual diversion as constrained by water availability. Program TABLES creates an optional vulnerability and resiliency table that includes the maximum monthly shortage, average sum of consecutive shortages, maximum number of consecutive shortages, and other shortage indices.

For new water right permits or amendments to existing permits, TCEQ criteria require that an agricultural irrigation right supply at least 75% of the proposed diversion target and at least 75% of the time computed on both a monthly and annual basis. Reliabilities of 100% are required for approval of new municipal water right permits. Existing reliabilities of senior rights are protected. Many older water rights do not meet the reliability criteria imposed on applicants for new or amended permits.

5. Brazos River Basin and Brazos Water Availability Model (WAM)

The monthly SIM or daily SIMD simulation model combined with an input dataset for the Brazos River Basin (Figures 1 and 2) and adjoining San Jacinto-Brazos Coastal Basin is called a Brazos WAM. Monthly Brazos WAM authorized and current use datasets are available at the TCEQ WAM website along with monthly datasets for all Texas river basins. A daily Brazos WAM authorized use scenario dataset available at the TAMU WRAP website reflects recently expanded modeling capabilities. A detailed technical report [34] documenting development of the daily Brazos WAM and investigation of various modeling issues is available at both the WRAP and TWRI websites. Daily Trinity and Neches WAM datasets and reports [35,36] can also be downloaded from the WRAP website [31]. Conversion of other monthly WAMs to daily are planned over the next several years.

5.1. Brazos River Basin and Adjoining Brazos-San Jacinto Coastal Basin

The Brazos River Basin encompasses an area of 119,000 square kilometers (km^2), with 111,000 km^2 in Texas and 8000 km^2 in New Mexico. The TCEQ WAM System combines the Brazos River Basin and adjoining San Jacinto-Brazos Coastal Basin in the same dataset. This coastal basin located south of the City of Houston between the Brazos and San Jacinto River Basins has a watershed area of 3000 km^2. Much of the water use from diversions from the Brazos River regulated by reservoirs shown in Figure 2 occur in the coastal plain south of Houston. Mean annual precipitation varies from 48 cm in the upper Brazos River Basin which lies in the high plains to 115 cm in the lower basin in the coastal region. The San Jacinto-Brazos Coastal Basin has a mean annual precipitation of 118 cm.

Mean daily observed flow rates at the U.S. Geological Survey (USGS) gauges near the cities of Waco and Richmond during January 1900 through July 2020 and October 1922 through July 2020 respectively, are plotted as Figures 3 and 4. The daily mean flows plotted in these figures reflect large long-term means but tremendous temporal variability in daily, monthly, and annual flows. The many water quantity and quality parameters included in the National Water Information System (NWIS) maintained by the U.S. Geological Survey (USGS) includes daily stream flows at 28,288 gauges, which include 1044 gauges in Texas [41]. Observed flows at 72 USGS gauges including the flows of Figures 3 and 4 were used in the compilation of naturalized stream flow data for the Brazos WAM.

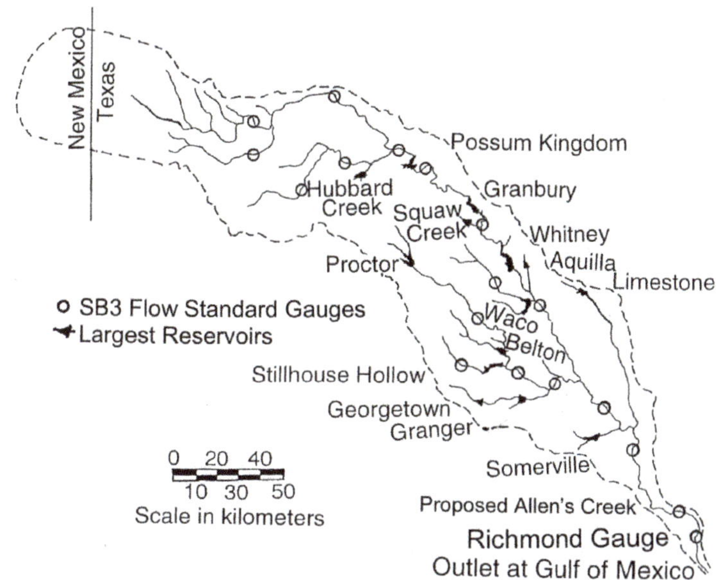

Figure 2. Sixteen largest reservoirs and nineteen gauge sites with SB3 EFS in the Brazos River Basin [27].

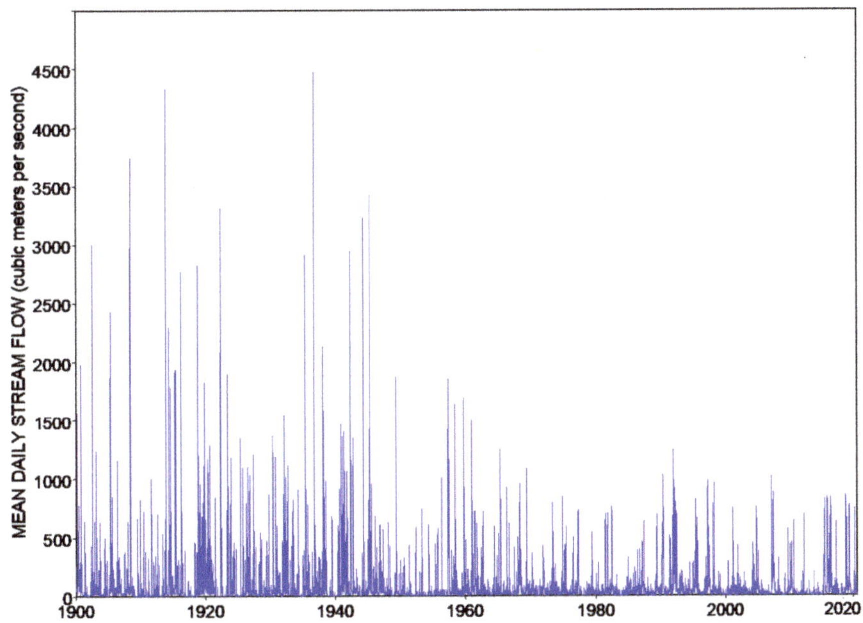

Figure 3. January 1900 through July 2020 daily flow of Brazos River at Waco gauge.

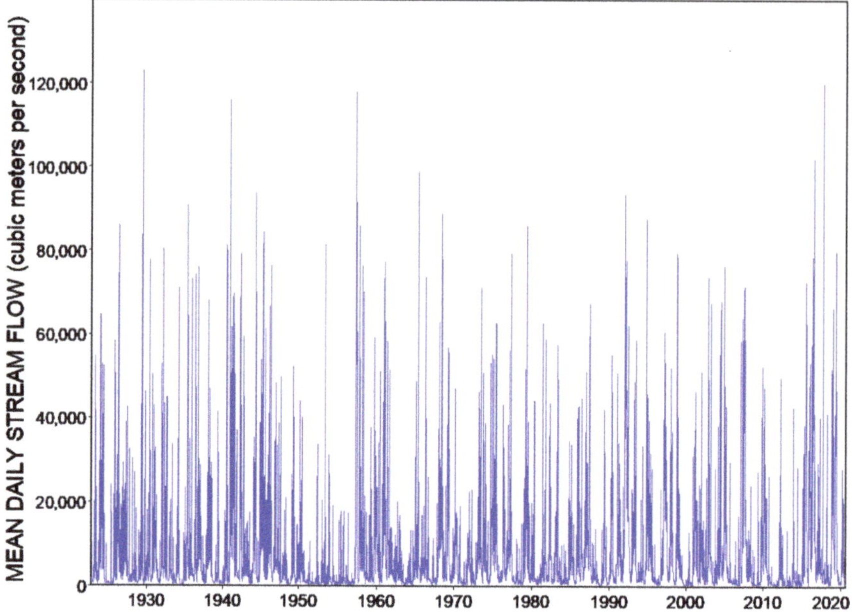

Figure 4. October 1922 through July 2020 daily flow of Brazos River at Richmond gauge.

5.2. Water Management in the Brazos River Basin and Adjoining Coastal Basin

The Brazos River Basin contains 673 reservoirs and the coastal basin has seven reservoirs cited in water right permits, of which 43 have conservation storage capacities of 6.17 million cubic meters or greater. The 16 reservoirs listed in Table 1 and included on the map of Figure 2 are the only reservoirs in the Brazos River Basin that have a combined conservation and flood control storage capacity of greater than 100 million cubic meters. There are no reservoirs this large in the coastal basin. These 16 reservoirs contain 80% of the total conservation storage capacity of the 680 reservoirs in the Brazos WAM and supply about 40% of the total annual permitted diversion volume.

The U.S. Army Corps of Engineers (USACE) owns and operates nine multiple-purpose reservoirs (Table 1) that contain all gated flood control storage capacity in the Brazos River Basin. Nonfederal sponsors control the storage capacity allocated to water supply and reimburse all costs allocated to water supply [42].

USACE flood control operations occur whenever lake levels rise above the top of the conservation pool and are based on non-damaging flow limits at downstream gauges. No releases are made that contribute to flows exceeding 708 cubic meters per second (m^3/s) at the Waco gauge, 1700 m^3/s at the Richmond gauge, or other specified non-damaging flow limits at other gauges. The effects of flood control operations of Whitney and Waco Reservoirs, with initial impoundment in 1951 and 1965 (Table 1), on flows at the gauge on the Brazos River near Waco are pronounced in Figure 3 because the gauge is located a short distance below the dams. The gauge on the Brazos River near Richmond is located significant distances downstream of all nine of the USACE flood control reservoirs. The effects of the dams are not as clearly evident in the flows at the Richmond gauge in Figure 4.

Water right permits authorize the use of stream flow to fill reservoir storage and supply water needs subject to specified conditions. Water right priorities reflecting the dates that stream flow was first appropriated or permit applications submitted range from 29 June 1914 to near the present for the Brazos River Basin and adjoining coastal basin. Over 1000 entities that include a river authority, water districts, cities, private companies, farmers, and other appropriators hold 1220 water right permits that authorize annual diversions totaling 3.05 billion m^3/year in the Brazos Basin and coastal basin for municipal (47.6%), industrial (30.1%), agricultural irrigation (18.0%), and other (4.3%) uses.

Table 1. Largest reservoirs in the Brazos River Basin.

Reservoir	Stream	Initial	Storage Capacity (Mm^3)		
		Impoundment	Conservation	Flood Control	Total
U.S. Army Corps of Engineers and Brazos River Authority					
Whitney	Brazos River	1951	785	1682	2467
Aquilla	Aquilla Creek	1983	65	115	180
Waco	Bosque River	1965	255	641	896
Proctor	Leon River	1963	73	388	462
Belton	Leon River	1954	565	790	1354
Stillhouse Hollow	Lampasas River	1968	291	487	778
Georgetown	San Gabriel	1980	46	116	161
Granger	San Gabriel	1980	81	220	301
Somerville	Yequa Creek	1967	198	428	626
Brazos River Authority					
Possum Kingdom	Brazos River	1941	894	–	894
Granbury	Brazos River	1969	191	–	191
Limestone	Navasota River	1978	278	–	278
Allen's Creek	Allen's Creek	proposed	180	–	180
City of Lubbock					
Alan Henry	Double Mountain	1993	143	–	143
West Central Texas Municipal Water District					
Hubbard Creek	Hubbard Creek	1962	392	–	392
Texas Utilities Services (cooling water for Comanche Peak Nuclear Power Plant)					
Squaw Creek	Squaw Creek	1977	187	–	187

The Brazos River Authority (BRA) has contracted for the conservation storage capacity in the nine federal reservoirs, owns three other existing reservoirs, and holds a water right permit for a proposed reservoir that is not yet constructed. The BRA also owns and operates regional water and wastewater treatment and water conveyance facilities. The BRA sells water under contract to cities, industries, and farmers subject to authorizations defined in the multiple water right permits held by the BRA. The City of Waco has multiple water right permits for Lake Waco, though the BRA is the nonfederal sponsor for the water supply storage in the federal reservoir. The BRA holds water right permits for the 11 other reservoirs of the 12-reservoir USACE/BRA system.

Hydroelectric energy is generated at Whitney Reservoir. Essentially all releases through the hydropower turbines are diverted downstream for municipal, industrial, or agricultural use. The conservation pool includes storage for head for hydropower as well as water supply. The electricity is marketed through a U.S. Department of Energy agency to a local electric power cooperative.

Environmental flow standards (EFS) have been established at the 19 USGS gauge sites on the Brazos River and its tributaries shown in Figure 2 following the process established pursuant to Senate Bill 3 (SB3) enacted by the Texas Legislature in 2007 [25,43]. An officially constituted expert science team [44] developed recommended EFS considering only environmental needs that were then refined by a stakeholder committee [45] based on consideration of all water needs. The science team and stakeholder committee submitted their recommendations to the TCEQ for final public and agency review, approval, and publication in the Texas Water Code [43]. The SB3 EFS include subsistence, base, and in-bank and overbank high-pulse flow components that vary seasonally and with hydrologic conditions. The procedures, reports, and other relevant information regarding establishment of SB3 EFS for the Brazos and other river basins are accessible at the TCEQ WAM website. The SB3 process for establishing EFS includes periodic review and improvement of the EFS.

5.3. *Water Availability Model (WAM) for the Brazos River Basin and Adjoining Coastal Basin*

The Brazos WAM simulates operation of 680 reservoirs and other facilities in accordance with 1220 water right permits. The authorized use scenario simulation with results presented in Section 5.4 is based on the premise that all water right holders appropriate the full amounts allowed in their water rights permits. Current use and other water use scenarios can also be simulated. The hydrologic period-of-analysis is January 1940 through December 2017. The 1940–2017 monthly naturalized flows at 77 control points provided in the simulation input dataset are disaggregated to daily and distributed to over 3000 other sites during the simulation.

SB3 EFS are incorporated the daily SIMD. Daily instream flow targets computed in a SIMD simulation in accordance with the EFS specifications are summed to monthly quantities within SIMD and recorded in a DSS file. The monthly targets are incorporated in the monthly SIM input dataset.

SB3 EFS are inserted in the WAM datasets with a priority based on the date that the designated science team and stakeholder committee submit recommendations to the TCEQ. The Brazos SB3 EFS were adopted in 2014 with a priority date of 1 March 2012. Existing senior water right permit holders are not affected. However, the SB3 EFS significantly reduce WAM simulated unappropriated flows available for future water right permit applicants.

The monthly SIM or daily SIMD simulation is based on the premise that water use requirements are supplied subject to water availability during each of the 936 months or 28,490 days of the 1940–2017 hydrologic period-of-analysis. The 1940–2017 naturalized flows provided as simulation input represent the stream flows that would have occurred naturally without human water resources development and use. Frequency and reliability metrics are computed from simulation results.

5.4. *Simulation Results*

SIM and SIMD simulation results can be massive. The modeling system provides flexible capabilities for organizing, analyzing, and displaying simulation results. Application of the modeling system in planning and administration of the water right permit process typically focuses on developing water supply reliability metrics for specific water rights of interest and assessing effects of these rights on the reliabilities of other water rights. Brazos WAM simulation results are used here in a more general basin-wide total manner to illustrate the concepts and issues discussed.

The mean, standard deviation, and quantities with specified exceedance frequencies (Equation (1)) for observed, naturalized, regulated, and unappropriated flows in cubic meters per second (m^3/s) at the Richmond gauge (Figure 2) for the 28,490 days or 936 months of the 1940–2017 hydrologic period-of-analysis are tabulated in Table 2. Metrics for daily and monthly means of observed and naturalized flows are tabulated in columns 2, 3, 4, and 5. Regulated and unappropriated flows computed alternatively in daily SIMD and monthly SIM simulations are compared in columns 6–11. SIMD simulation results include both simulated daily (columns 6 and 8) and aggregated monthly (columns 7 and 9) quantities. Statistics for monthly SIM results are presented in columns 10 and 11.

The characteristics of observed versus naturalized versus simulated regulated flows of the Brazos River at the Richmond gauge site (Figure 2) are reflected in the statistics of Table 2. The 1220 water rights in the Brazos WAM reduce the mean flow of 228 m^3/s at the basin outlet for natural undeveloped conditions to 181 m^3/s for the simulated scenario of all water rights appropriating their authorized amounts. The frequency statistics indicate that unappropriated flows can be expected to be zero much of the time, which implies that significant reservoir storage capability is required to achieve acceptable levels of water supply reliability for additional new or increased water rights. The averaging effects of monthly versus daily computational time steps can also be observed in Table 2.

Table 2. Frequency statistics for daily and monthly observed and naturalized flows at the Richmond gauge site and regulated and unappropriated flows from daily and monthly simulations.

1	2	3	4	5	6	7	8	9	10	11
	Observed		Naturalized		Daily SIMD Simulation				Monthly SIM	
					Regulated		Unappropriated		Regulated	Unappro-priated
	Daily (m^3/s)	Monthly (m^3/s)	Daily (m^3/s)	Monthly (m^3/s)	Daily (m^3/s)	Monthly (m^3/s)	Daily (m^3/s)	Monthly (m^3/s)	Monthly (m^3/s)	(m^3/s)
Mean	216	216	228	228	181	181	103	103	180	113
Standard Deviation	346	273	407	292	369	250	289	192	266	238
Minimum	1.6	5.7	0	0	0	0	0	0	0.1	0
99%	8.7	12.2	4	7	0	0.1	0	0	4.4	0
98%	11	14	6.7	9.6	0	2.6	0	0	6.8	0
95%	14.8	18	11	15.9	0	8	0	0	10.3	0
90%	19.8	24.5	16.5	24.1	1.5	12.9	0	0	15.2	0
80%	29.2	35.8	27.5	38.3	10.4	20.4	0	0	22.4	0
70%	39.9	52.3	39.9	59.1	19.2	31.7	0	0.5	31.9	0
60%	54.7	74.8	57.2	81.9	25.9	49	0	3.6	42.8	0
50%	80.7	105	83.8	120	41.4	78	0	14.7	68.9	0
40%	123	156	127	168	75.2	119	0	35.8	108	19.2
30%	189	222	195	237	140	180	21.6	79.2	171	65.9
20%	309	358	309	364	248	300	110	170	283	184
10%	555	563	582	579	499	509	312	329	506	370
Maximum	3400	2190	9200	2880	9170	1810	7300	1480	2560	2530

Daily SB3 EFS instream flow targets at the Richmond gauge near the outlet computed in the daily SIMD simulation are plotted in Figure 5. The monthly SIM simulation of Table 2 and Figure 6 includes SB3 EFS flow targets from the daily SIMD simulation for the 19 sites shown in Figure 2 computed as a function of regulated flow, season of the year, and hydrologic condition, as specified by the SB3 EFS with subsistence, base, and high-pulse flow components [25,34,43]. The SB3 EFS do not affect water rights with seniority dates earlier than 1 March 2012. In the model, junior rights with priority dates later than this date curtail actions that adversely affect meeting the requirements defined by the SB3 EFS.

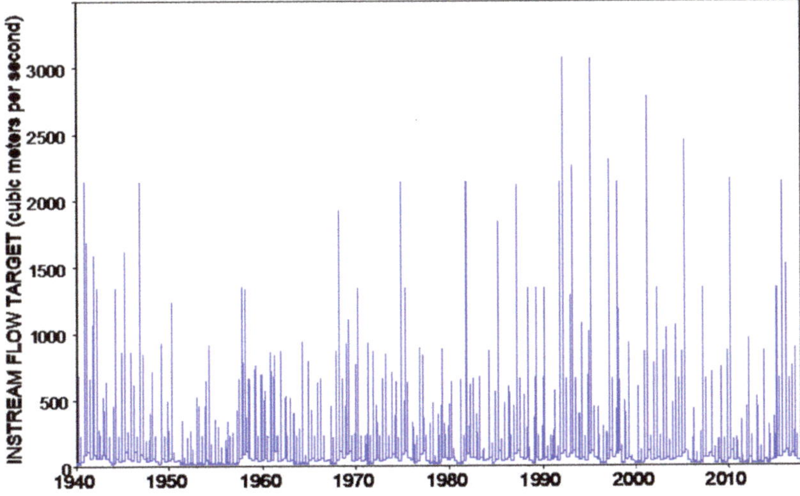

Figure 5. Daily minimum instream flow targets for SB3 EFS at Richmond gauge on the Brazos River.

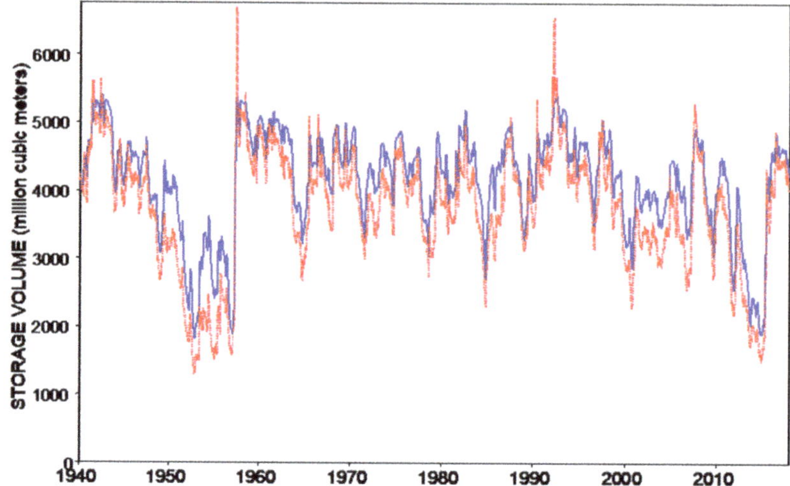

Figure 6. Summation of total storage contents of the 680 reservoirs in the monthly SIM (blue solid line) and daily SIMD (red dotted line) simulations.

Summations of SIMD end-of-day and SIM end-of-month storage contents of the 680 reservoirs in the Brazos WAM from daily and monthly simulations are plotted in Figure 6. These plots reflect operation of the 680 reservoirs, most of which were constructed during the 1950s through 1980s, to supply water use targets authorized by the 1220 presently active water right permits during an assumed repetition of 1940–2017 hydrology. Storage in individual reservoirs is of interest in most applications and tends to fluctuate much more than the total storage in 680 reservoirs.

Reservoir storage contents provide a meaningful drought index. The most hydrologically severe drought in the Brazos River Basin since before 1940 began gradually in 1950 and ended with major widespread flooding in April 1957, as shown in Figure 6. The more economically costly 2010–2014 drought and other less-severe dry periods are also evident in the storage plots. The residents of the Brazos River Basin, and most other areas of Texas, and the water management community have never experienced a drought as hydrologically severe as 1950–1957 with present population, economic development, and associated water needs. Water planning and management is based on a drought more hydrologically severe than 1950–1957 occurring at some unknown time in the future.

The 1940–2017 mean annual natural flow of the Brazos River near its outlet is 236% of the annual diversions, totaling 3.05 billion cubic meters per year authorized by the 1220 water right permits modeled in the Brazos WAM. The majority of the flow occurs during periods of high flows or floods. Reservoir storage is essential for reliable water supplies. The volume reliability, R_V in Equation (3), for the 3.05 billion m^3/year aggregation of all water supply diversion rights authorized by the 1220 water right permits are 79.1% and 87.6% respectively, in the daily and monthly simulations.

6. Hydrologic and Institutional Aspects of Water Management and Modeling Thereof

Important considerations and issues encountered in assessing water availability and supply reliability statewide and allocating stream flow and reservoir storage among numerous water users and diverse types of use are highlighted as follows. The Brazos WAM serves as an example to illustrate key concepts and issues. Two distinctly different but integrally interconnected topics are addressed: (1) water management and (2) modeling and analysis of water management.

6.1. Hydrologic Variability and Stationarity

Variability and stationarity of precipitation, reservoir evaporation, and stream flow are key considerations affecting water management and assessments of water availability. Hydrologic variability includes continuous fluctuations and seasonal changes along with the extremes of intense floods and severe multiple-year droughts. Hydrologic variability and associated water supply reliability, flood risk, and future uncertainty are fundamental to water management and modeling thereof. Stationarity, or lack thereof (non-stationarity), refers to long-term homogeneity over time with no permanent changes or trends. Stationarity of naturalized stream flows and other variables is also important in water availability modeling and water management.

The TWDB maintains a database updated annually of January 1940 to near-present mean monthly precipitation rates and January 1954 to near-present monthly reservoir water surface evaporation rates for each of 92 one-degree quadrangles that encompass the state [4,26]. The databases are used along with data from other sources to develop simulation input datasets of net reservoir evaporation less precipitation rates for the WAMs. Evaporation–precipitation volumes are computed in the simulation model by multiplying fluctuating reservoir surface areas by evaporation less adjusted precipitation rates which exhibit year-to-year as well as great seasonal variability.

Evaporation is a major component of reservoir water budgets and important consideration in water management and water availability assessments. For comparison, the simulated long-term mean annual evaporation volume from the over 3400 reservoirs statewide has been computed with the WAMs to be a volume equivalent to 61% of the year 2010 actual annual total agricultural or 126% of the total municipal water use from all surface and groundwater sources in Texas [46].

The WRAP program HYD includes routines for managing the TWDB precipitation and reservoir evaporation rate datasets and performing statistical frequency and trend analyses of the data for individual quadrangles and statewide averages [4]. Long-term trends or permanent changes in 1940–2019 precipitation or 1954–2019 evaporation characteristics are not evident from time series plots and regression analyses of the 92 TWDB datasets reflecting spatial averaging over one-degree longitude by one-degree latitude quadrangle areas. Any long-term trends that may exist are hidden by the great continuous variability. Statewide averages of monthly precipitation and reservoir water surface evaporation rates in centimeters (cm) per month are plotted in Figures 7 and 8.

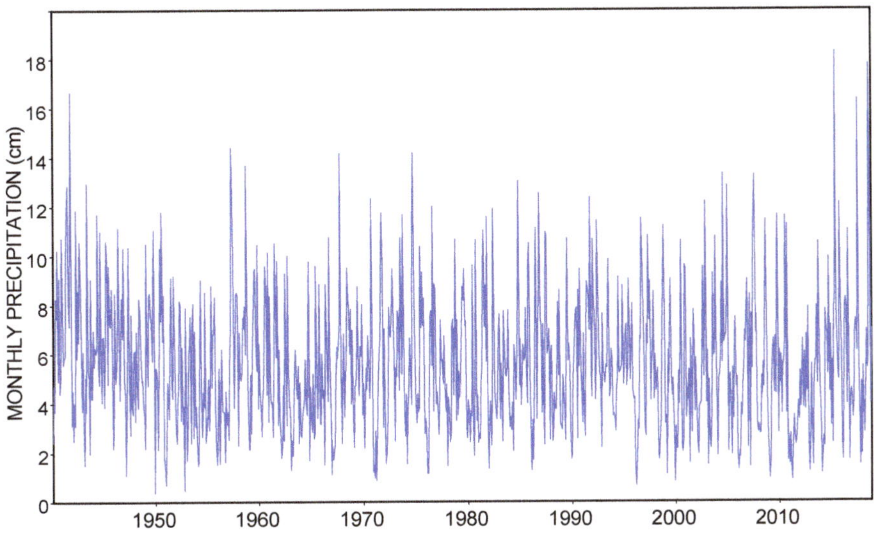

Figure 7. Statewide average monthly precipitation rates.

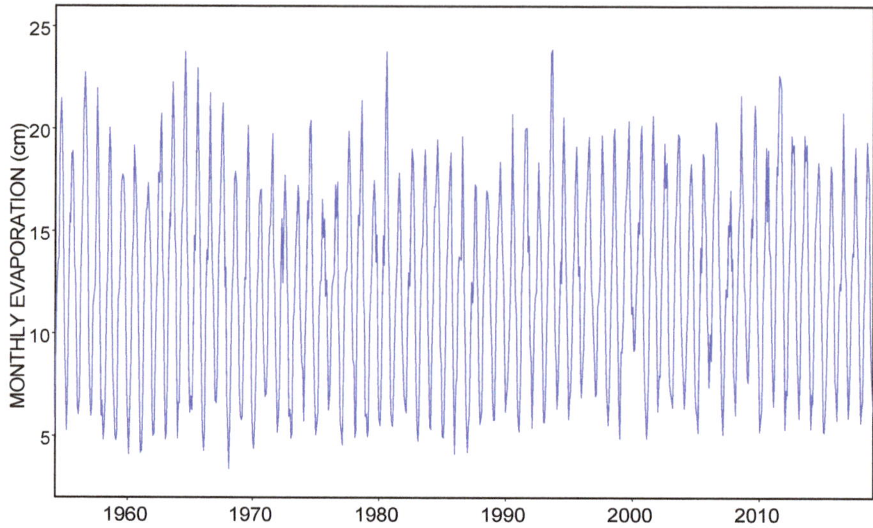

Figure 8. Statewide average monthly reservoir evaporation rates.

The observed daily flows of the Brazos River at the USGS gauges near Waco and Richmond plotted in Figures 3 and 4 illustrate the tremendous variability of river flows throughout Texas, including throughout the Brazos River Basin. Extremes of multiple-year droughts and major floods are combined with seasonal and continuous fluctuations.

Construction and operation of dams and other facilities, water supply diversions, return flows from surface and groundwater supply sources, and other aspects of population and economic growth significantly affect stream flow, with the resulting changes varying greatly between locations [26]. For example, the flows of the San Antonio River below the City of San Antonio and the San Jacinto River below Houston have increased significantly over the past 100 years as a result of wastewater treatment plant effluent accompanying increased water supply from groundwater and inter-basin conveyance and increase impervious land cover due to urbanization. The flow of the Rio Grande has deceased greatly due to construction of reservoirs and development of irrigated agriculture. The Brazos and Trinity Rivers are representative of many rivers that have experienced a decrease in flood flows due to flood control reservoirs and raising of low flows due to return flows from municipal and industrial water use. Flow immediately below dams is greatly affected by reservoir operations, but the effects diminish with distance downstream, as illustrated by Figures 3 and 4.

The WRAP/WAM modeling process consists of computational adjustments that convert observed flows to naturalized flows input to a simulation model that generates regulated and unappropriated flows reflecting a specified scenario of water resources development, allocation, management, and use. The process of naturalizing flows consists of removing non-stationarities. Removal of all non-stationarities is not feasible. However, the Texas experience in developing the WAMs indicates that long-term changes in flow characteristics are due primarily to major reservoir projects and major water supply diversions and return flows, which are included in the flow naturalization adjustments adopted in compilation of the WAM simulation input datasets. Based on statistical trend analyses and time series plots, the naturalized flows in the WAMs representing past natural conditions are considered to generally be reasonably free of long-term changes or trends.

Effects of long-term future climate change on hydrology and water management throughout the world are explored extensively in the literature. The Brazos WAM and San Jacinto River Basin WAM were combined with global climate model precipitation and evaporation output and a watershed precipitation-runoff model in university research studies to evaluate the impacts of future climate change scenarios on water availability [47,48]. Modeling uncertainties were found to be too great to derive meaningful conclusions regarding future climatic and hydrologic conditions in these studies. Neilson-Gammon et al. [49] assess the risk and consequences of unprecedented future drought conditions occurring in Texas during the latter half of the 21st century due to climate change.

6.2. Water Management Community

Assessments of water availability are performed within a complex water management community of diverse entities with different responsibilities and roles. WRAP and its input datasets in the WAM system were developed and are employed within the Texas water management community.

With over 3000 employees, the TCEQ is the largest state environmental regulatory agency in the U.S. Along with its many other responsibilities, the TCEQ administers five interstate river basin compacts and two water right permit systems for (1) the Texas share of the waters of the Rio Grande and (2) the remainder of Texas. The TCEQ leadership role in developing, maintaining, and expanding the WAM system stems from its water allocation responsibilities.

Both a regional and statewide planning process and creation of the WAM system were authorized by comprehensive water management legislation enacted in 1997 as Senate Bill 1 and now commonly referenced as SB1. Sixteen regional plans and a statewide plan updated in a five-year cycle forecasts water needs and water availability at 10-year intervals for 50 years into the future and presents plans for dealing with deficits. The TWDB in collaboration with regional planning groups is responsible for SB1 regional and statewide planning and assists local water supply entities in financing water projects. TCEQ approval of applications for new water right permits or amendments to existing permits requires that proposed actions be consistent with SB1 statewide and relevant regional water plans. The shared WAM system contributes significantly to integration of planning and water allocation.

The TWDB manages the Texas Instream Flow Program (TIFP) authorized by the 2001 SB2 to improve capabilities for preserving stream flows for environmental needs. Recognizing that many years will be required to develop all of the needed assessment methods and water management strategies, the 2007 SB3 initiated procedures for incorporating EFS in the WAMs based on best currently available expert opinion and information, subject to continuing review and improvement. A science team and stakeholder committee in collaboration with the TCEQ established the SB3 ESF for the Brazos River Basin following SB3 protocols [43–45]. Science teams are comprised of hydrologists and ecological scientists from universities, consulting firms, and government agencies. Stakeholder committees are constituted to represent a diverse range of interests that include municipal, industrial, and agricultural water users, electric utilities, recreation, environmental protection, and other relevant sectors.

The Brazos River Authority (BRA), created in 1931, is the oldest of the 19 Texas river authorities and has water management responsibilities for a river basin with an area larger than many states in the U.S. and countries in the world. The 19 river authorities of Texas were created by the Texas Legislature. They are funded primarily through their sale of water supply services and electricity to other public and private entities. River authorities hold many of the water rights that include larger storage and diversion quantities. Unlike the TCEQ and TWDB, the river authorities own and operate reservoir projects, water treatment and conveyance facilities, and other constructed infrastructure.

Many cities and private entities hold their own water right permits issued by the TCEQ. Other cities, private companies, and farmers purchase water from river authorities or water districts that hold the required TCEQ-administered water right permits. Some larger cities supply neighboring smaller cities. Municipal water districts are created through cooperative agreements of multiple cities. Farmers may purchase water from irrigation districts. The numerous water districts are similar to river authorities but have more narrowly defined responsibilities.

The U.S. Congress in the Flood Control Act of 1936 charged the U.S. Army Corps of Engineers (USACE) with construction and operation of flood control projects nationwide at federal expense. The USACE is also responsible for inland navigation. Water supply is a local responsibility. The Water Supply Act of 1958 authorized inclusion of water supply storage in multiple-purpose federal reservoirs subject to all costs allocated to water supply being reimbursed to the federal government by nonfederal sponsors [42]. The USACE owns and operates over 500 reservoirs nationwide. Nine of the 27 USACE reservoirs in Texas are located in the Brazos River Basin. The USACE contracts with nonfederal sponsors that control the portion of reservoir storage capacity allocated to water supply but provides no commitment regarding the availability of water to fill the storage capacity. The USACE is not directly involved with obtaining or administering water rights.

The USACE also administers a permit program under authority of Section 404 of the Clean Water Act of 1977 regulating construction activities affecting rivers, streams, and wetlands. The use of the WRAP/WAM system to evaluate Section 404 permit applications for construction of water supply projects in Texas is being investigated by the USACE Fort Worth and Galveston District Offices.

Water right permit applicants, regional planning groups, and various other entities routinely hire consulting engineering firms to perform professional services that include WRAP/WAM simulation studies. The many consulting firms that have employed WAMs for various clients range in size from firms consisting of one professional engineer to regional firms with staff of several hundred professionals working in offices in multiple Texas cities to international companies, with many thousands of people distributed between many different offices in Texas and throughout the world.

The Water Resources Act of 1964 authorized establishment of a water institute at a university in each state to facilitate federal/state partnerships in research and extension. These state institutes comprise the National Institutes for Water Resources (NIWR) network managed by the U.S. Geological Survey (USGS) at the federal level. The Texas Water Resources Institute (TWRI) of the Texas A&M University System represents Texas in the NIWR network. The WRAP modeling system originated from a university research project sponsored by this federal/state partnership program.

The membership of the Texas Water Conservation Association (TWCA) is comprised of water management professionals employed by the many public and private entities mentioned in the preceding paragraphs. The WRAP Committee of the TWCA provides recommendations to the TCEQ and its contractor (TAMU represented by this author) regarding water management issues and needs for expanded modeling and analysis capabilities and reviews research and development products. Eleven WRAP user group conferences held since 2006 have been attended by water professionals from the TCEQ, TWDB, river authorities, water districts, other state and federal agencies, engineering firms, and universities. WRAP training sessions are conducted periodically.

The author was the recipient of the Research and Innovation Award of the American Academy of Water Resources Engineers (AAWRE) presented at the 20th American Society of Civil Engineers (ASCE) Environmental and Water Resources Institute (EWRI) World Water Congress in 2019 for his role in development of WRAP and its implementation in the Texas WAM System.

6.3. Water Allocation

As demands on limited resources intensify, water allocation through water right permit systems, interstate compacts, international treaties, federal/state/local agreements, and environmental protection programs grows in importance and significantly affects water availability. The WRAP modeling system includes flexible features for simulating diverse water allocation mechanisms.

Water allocation systems equitably apportion water among users, protect existing water users from having their supplies diminished by new users, govern the sharing of limited water resources during droughts when supplies are inadequate to meet all needs, and facilitate efficient use of water resources. Each of the 50 states in the U.S. has developed its own rules and practices, which have evolved historically and continue to change [42,43]. Western and eastern states have generally adopted different approaches to water rights due largely to the western states having much drier climates. Most states treat allocation of groundwater versus surface water very differently.

Water flowing in the Rio Grande and stored in International Amistad and Falcon Reservoirs on the Rio Grande are jointly controlled by the Mexico and U.S. Sections of the International Boundary and Water Commission (IBWC). Flow and storage are allocated between the two nations by a 1944 treaty.

The IBWC maintains an accounting of storage in Amistad and Falcon Reservoirs, inflows, water supply and hydropower releases, spills, and evaporation allocated to Mexico and to the U.S. Texas participates in interstate river basin compacts with New Mexico, Oklahoma, Arkansas, and Louisiana for allocation of the water resources of the Rio Grande, Pecos, Canadian, Red, and Sabine Rivers. The WAMs simulate allocation of water between Texas and its neighbors, allocate the Texas share to all individual water rights in Texas, but do not further sub-allocate the water allocated to the other states and Mexico.

Legal rights to the use of stream flow in the U.S. are generally based on two alternative doctrines, riparian and prior appropriation [43]. The basic concept of the riparian doctrine is that water rights are incidental to the ownership of land adjacent to a stream. The prior appropriation doctrine is based on protecting senior water users from having their supplies diminished by newcomers developing water supplies later in time. In a prior appropriation system, rights are not inherent in land ownership, and priorities are established based on dates that water is appropriated.

Variations of the riparian doctrine are applied in 29 states in the eastern and central U.S. The prior appropriation doctrine governs water rights in 19 western states, including Texas. Ten of these states, including Texas, originally recognized riparian rights but later converted to prior appropriation while preserving existing riparian rights. Hawaii and Louisiana have their own unique water right systems. Most of the western states have established permit systems in which a state agency issues permits to water right holders specifying amounts and conditions of water use. With growing demands on limited water resources, permit systems will likely continue to be developed in the eastern states similar to those already in place in the drier western states [22,50,51].

Several western states have water-master operations for real-time management of water rights, but most states do not. The TCEQ Rio Grande Water Master Office has administered accounting of water use, working closely with irrigators, cities, and the IBWC, since the 1970s. The Brazos River Basin water-master office was established in 2015. However, water master operations have not yet been established for the majority of Texas river basins. The TCEQ administers curtailment actions during drought and takes enforcement action any time to stop reported unauthorized water use but does not otherwise closely monitor water use. Establishment of additional TCEQ water master offices for individual river basins with more detailed monitoring and accounting procedures continues to be investigated.

With the exception of the Rio Grande, water allocation priorities are set by dates specified in water right permits that reflect dates water was initially appropriated. Priorities for flow and storage in the lower Rio Grande are based on type of use as well as historical use. Modeling priority systems is an essential fundamental requirement for the WAMs. WRAP includes flexible options for both simulating variations of the prior appropriation water rights doctrine and alternatively appropriating water in an upstream-to-downstream sequence consistent with the riparian rights doctrine.

6.4. Reservoir System Operations

A Brazos River Authority (BRA) system operation permit with accompanying water management plan approved by the TCEQ in November 2016 significantly increased water supply capabilities based on an expanded understanding of reliability provided by the WRAP/WAM modeling system. The amount of water that BRA supplies under contracts with wholesale water customers is constrained by its water right permits. BRA water rights were established historically for individual reservoir projects near the time of their construction. The new system operations permit credits the BRA with using unregulated flow entering the river system below the dams and return flows from BRA wastewater treatment plants in coordination with releases from eleven reservoirs that balance storage between the reservoirs.

One key basic concept of the system operation permit and water management plan is that for a particular level of reliability, the total quantity of water provided by multiple reservoirs operated as a system is greater than the summation of quantities provided by the reservoirs with each operated individually. Storage contents can be balanced in multiple reservoirs to minimize the risk that any one reservoir is emptied and thus unable to supply demands. The hydrologic characteristics of large river basins include spatial variability of the timing of low flow conditions at different locations.

Another key system operations concept is to execute water supply contracts that commit different levels of reliability, called firm and interruptible, for different types of water use and available alternative water supply sources. Municipal water supplies require a high level of reliability. Farmers may prefer to increase the amount of water normally available in many years for irrigation even though the risk of shortages during drought years increase. Declining groundwater sources limit groundwater use. However, infrequent increased use of groundwater can be combined with commitments for increased surface water use most of the time.

For example, the Lower Colorado River Authority (LCRA) operates a system of six reservoirs on the Colorado River to supply water for Austin and other cities for municipal and industrial use at a high level of reliability and water to farmers in the lower basin for agricultural irrigation at significantly lower levels of reliability. LCRA agreements with farmers for irrigation water are based on setting allocations at the beginning of the annual irrigation season based on the storage contents of the reservoirs. The irrigators received no water during the extremely dry 2011. This water allocation strategy and most of the reservoirs did not exist during the historic 1950–1957 drought.

The older LCRA and recent BRA system operation permits and water management plans reflect the tradeoffs that occur between the amount of water committed for beneficial use and the level of reliability that can be achieved. If water commitments are limited as required to ensure an extremely high level of dependability, much of the water resource flows to the ocean or is lost through reservoir evaporation much of the time. WAM studies in the various river basins indicate that quantities that may be supplied change greatly with relatively small changes in reliability requirements. The amount of water supplied from Texas river systems can be increased significantly by accepting higher risks of shortages or emergency demand reductions.

Reuse of returns flows is another important system operations consideration. The BRA system operation permit application process included extensive public review and comment. Several cities expressed concerns that BRA was claiming their wastewater treatment effluent. The final approved permit credits the BRA with reuse of only return flows from its own regional wastewater treatment plants. The WAMs have also been applied in exploring the effects of access by different entities to wastewater treatment plant return flows in the City of Austin on the Colorado River and the Dallas and Fort Worth metropolitan area in the upper Trinity River Basin.

6.5. Major Limitations of the Modeling System

Complexities discussed throughout this paper impose limitations on both computer modeling and water management. Several diverse constraints on modeling capabilities for assessing water availability and allocation are highlighted as follows.

Stream flow is extremely variable. A monthly computational time step has been concluded to be optimal for modeling water allocation and management from the perspective of municipal, industrial, agricultural, and other types of water use. However, daily computations are required to adequately capture the effects of flow variability from the perspectives of flood control reservoir operations and environmental instream flow standards, particularly high-pulse flow components of the standards. As noted in Section 4.2, flow routing and forecasting are employed with a daily model, though not relevant in a monthly model. Flow routing and forecasting are highly approximate. Calibration of routing parameters for stream reaches has been found to be complex and inaccurate.

Losses of flow in river reaches due to seepage and evapotranspiration are considered in the downstream translation of flow changes due to water right actions in both the monthly and daily versions of the model. The loss computation methodology is simplistic due to difficulties in both simulating the relevant physical processes and determining values for input parameters.

Conjunctive management of surface and ground water, or lack thereof, is an important issue in both water management and modeling thereof. Fundamental hydrologic and institutional differences prevent combining surface and ground water in the same simulation model. Unlike surface water, groundwater ownership is inherent in land ownership. Conjunctive management of water from ground and surface water is constrained by differences in allocation policies. Physical hydrologic processes and fundamental modeling strategies and methods are also very different between these interconnected components of the hydrologic cycle.

Water availability for beneficial use depends upon water quality as well as quantity. For example, the water supply capabilities of several large reservoirs in Texas are severely constrained by salinity from natural salt deposits in geologic formations in the Permian Basin geologic region that underlies the upper watersheds of the Rio Grande, Pecos, Brazos, Red, and Canadian River Basins in New Mexico, Oklahoma, and the north Texas panhandle. Stream flows have high concentrations of chlorides, sulfates, and other dissolved minerals in the upper reaches of these river systems that are diluted by low-salinity tributary inflows in the middle and lower basins. A salinity simulation component of the WRAP modeling system has been developed motivated by the natural salt pollution and applied with the monthly Brazos WAM to explore effects of salinity on water supply capabilities [6,39]. However, much more research is needed to improve capabilities for assessing the impacts of natural salt pollution and other water quality issues on water availability and allocation for beneficial use.

A pure prior appropriation water rights system is not feasible for many reasons. For example, although the WRAP simulation model allows reservoir storage and water supply diversions to be assigned different priorities, in most water right permits, a single priority date is assigned in a permit granting the right to both store and divert water. Reservoir operation in Texas is based on long-term storage as a protection against severe multiple-year droughts. The supply reliability of a reservoir is diminished if upstream junior appropriators reduce inflows when the reservoir is not completely full and spilling. However, forcing junior appropriators to curtail their water use to maintain inflows to an almost full or even significantly drawn-down reservoir is difficult and not necessarily the optimal use of the water resource. This is an example of a water policy issue that is difficult to resolve though potential solution strategies can be easily simulated in the model.

7. Conclusions

Quantitative probability-based assessments of water availability are essential for effective water allocation and management. Modeling of institutional mechanisms as well as river system hydrology and operation of dams/reservoirs and other constructed facilities are necessary in assessments of water availability. Successful implementation of the Texas WAM System required collaborative efforts of a large and diverse water management community. The shared use of the modeling system has significantly contributed to integrating water allocation, planning at statewide, regional, project feasibility, and operational levels, research and development, and other water management endeavors.

Assessments of water availability and supply reliability are performed with the WRAP/WAM system in three stages: (1) compilation and continuing updating of simulation input datasets, (2) performing simulations, and (3) organizing and analyzing relevant frequency and reliability metrics and other information from the simulation results. Water availability assessment applications usually involve revising simulation input datasets to reflect changes in water use requirements or different proposed projects or management strategies of interest. The simulation model combines extremely variability natural river system hydrology, complex operations of constructed infrastructure, and water allocation systems that grow in importance with increasing demands on limited resources.

The generalized Water Rights Analysis Package (WRAP) modeling system is applicable in any place in the world and reflects flexibility and practicality necessitated by its evolution within the Texas water management community. Lessons learned from the Texas experience in creating and employing a water availability modeling system are relevant worldwide.

Funding: This research received no external funding.

Acknowledgments: Many agencies and professionals in the Texas water management community have contributed to the evolution of the WRAP/WAM modeling system over the past 30 years. However, the information and conclusions presented in the paper are the responsibility of the author without implication of official endorsement by any of these agencies or individuals.

Conflicts of Interest: The author declares no conflict of interest.

References

1. Wurbs, R.A. *Water Rights Analysis Package Modeling System Reference Manual*, 12th ed.; Technical Report (TR) 255; Texas Water Resources Institute: College Station, TX, USA, 2019; 462p. Available online: https://twri.tamu.edu/publications/technical-reports/2019-technical-reports/tr-255/ (accessed on 1 October 2020).
2. Wurbs, R.A. *Water Rights Analysis Package Modeling System Users Manual*, 12th ed.; TR-256; Texas Water Resources Institute: College Station, TX, USA, 2019; 272p. Available online: https://twri.tamu.edu/publications/technical-reports/2019-technical-reports/tr-256/ (accessed on 1 October 2020).
3. Wurbs, R.A. *Fundamentals of Water Availability Modeling with WRAP*, 9th ed.; TR-283; Texas Water Resources Institute: College Station, TX, USA, 2019; 114p. Available online: https://twri.tamu.edu/publications/technical-reports/2019-technical-reports/tr-283/ (accessed on 1 October 2020).
4. Wurbs, R.A. *Water Rights Analysis Package River System Hydrology*, 3rd ed.; TR-431; Texas Water Resources Institute: College Station, TX, USA, 2019; 215p. Available online: https://twri.tamu.edu/publications/technical-reports/2019-technical-reports/tr-431/ (accessed on 1 October 2020).
5. Wurbs, R.A.; Hoffpauir, R.J. *Water Rights Analysis Package Daily Modeling System*; TR-430; Texas Water Resources Institute: College Station, TX, USA, 2019; 430p. Available online: https://twri.tamu.edu/publications/technical-reports/2019-technical-reports/tr-430/ (accessed on 1 October 2020).
6. Wurbs, R.A. *Salinity Simulation with WRAP*; TR-317; Texas Water Resources Institute: College Station, TX, USA, 2009; 87p.
7. Wurbs, R.A. Texas water availability modeling system. *J. Water Resour. Plan. Manag.* **2009**, *131*, 270–279. [CrossRef]
8. Wurbs, R.A. Sustainable statewide water resources management in Texas. *J. Water Resour. Plan. Manag.* **2015**, *141*, A4014002. [CrossRef]
9. Wurbs, R.A. *Water Management Models*; Pearson/Prentice-Hall Publishing: Upper Saddle River, NJ, USA, 1995; 245p.
10. Wurbs, R.A. Dissemination of generalized water resources models in the United States. *Water Int.* **1998**, *23*, 190–198. [CrossRef]
11. Wurbs, R.A. *Modeling and Analysis of Reservoir System Operations*; Pearson/Prentice-Hall: Upper Saddle River, NJ, USA, 1996; 372p.
12. Labadie, J.W. Optimal operation of multireservoir systems: State-of-art review. *J. Water Resour. Plan. Manag.* **2004**, *130*, 93–111. [CrossRef]
13. Rani, R.; Moreira, M.M. Simulation-optimization modeling: A survey and potential applications in reservoir system operation. *Water Resour. Manag.* **2010**, *24*, 1107–1138. [CrossRef]
14. Lund, J.R.; Hui, R.; Escriva-Bou, A.; Porse, E.; Adams, L.; Connaughton, J.; Kasuri, L.; Lord, B.; Siegfried, L.; Thayer, R.; et al. Chapter 130 Reservoir operation design. In *Handbook of Applied Hydrology*; Singh, V.P., Ed.; McGraw Hill: New York, NY, USA, 2017.
15. Wurbs, R.A. *Comparative Evaluation of Generalized River/Reservoir System Models*; TR-282; Texas Water Resources Institute (TWRI): College Station, TX, USA, 2005; 203p.
16. Wurbs, R.A. Chapter 1 Generalized models of river system development and management. In *Current Issues in Water Management*; Uhlig, U., Ed.; InTech: Rijeka, Croatia, 2011; pp. 1–22. Available online: https://www.intechopen.com/books/current-issues-of-water-management/generalized-models-of-river-system-development-and-management (accessed on 1 October 2020).
17. U.S. Army Corps of Engineers Hydrologic Engineering Center. HEC-ResSim. Available online: https://www.hec.usace.army.mil/software/hec-ressim/ (accessed on 1 October 2020).
18. Center for Advances Decision Support for Water and Environmental Systems. RiverWare. Available online: https://www.riverware.org/ (accessed on 1 October 2020).

19. Colorado State University. MODSIM Decision Support System. Available online: http://modsim.engr.colostate.edu/ (accessed on 1 October 2020).
20. Texas Water Development Board. *Water for Texas 2017*; Texas Water Development Board: Austin, TX, USA, 2017.
21. Sahs, M.K. (Ed.) *Essentials of Texas Water Resources*, 6th ed.; State Bar of Texas, Environmental and Natural Resources Law Section: Austin, TX, USA, 2020.
22. Wurbs, R.A. Water rights in Texas. *J. Water Resour. Plan. Manag.* **1995**, *121*, 447–454. [CrossRef]
23. Texas Water Development Board Planning. Available online: https://www.twdb.texas.gov/waterplanning/index.asp (accessed on 1 October 2020).
24. National Research Council. *The Science of Instream Flows: A Review of the Texas Instream Flow Program*; National Academies Press: Washington, DC, USA, 2005.
25. Wurbs, R.A. Incorporation of environmental flows in water allocation in Texas. *Water Int. Int. Water Resour. Assoc.* **2017**, *42*, 18–33. [CrossRef]
26. Wurbs, R.A.; Zhang, Y. *River System Hydrology in Texas*; TR-461; TWRI: College Station, TX, USA, 2014; 442p. Available online: https://twri.tamu.edu/publications/technical-reports/2014-technical-reports/tr-461/ (accessed on 1 October 2020).
27. Gippel, C.J.; Cosier, M.; Markar, S.; Liu, C. Balancing environmental flows needs and water supply reliability. *Int. J. Water Resour. Dev.* **2009**, *25*, 331–353. [CrossRef]
28. O'Keefe, J. Chapter 4 Environmental Flow Allocation as a Practical Aspect of Integrated Water Resources management. In *River Conservation and Management*; Boon, P.J., Raven, P.J., Eds.; John Wiley & Sons: Hoboken, NJ, USA, 2012.
29. Jain, S.K.; Acreman, M.C. Chapter 134 Environmental flows. In *Handbook of Applied Hydrology*; Singh, V.P., Ed.; McGraw-Hill: New York, NY, USA, 2016.
30. Wohl, E. Forgotten legacies: Understanding and mitigating historical human alterations of river corridors. *Water Resour. Res.* **2019**, *55*, 181–201. [CrossRef]
31. Water Rights Analysis Package. Available online: https://ceprofs.civil.tamu.edu/rwurbs/wrap.htm (accessed on 1 October 2020).
32. Texas Water Resources Institute Publications. Available online: http://twri.tamu.edu/publications/ (accessed on 1 October 2020).
33. Wurbs, R.A.; Walls, W.B. Water rights modeling and analysis. *J. Water Resour. Plan. Manag.* **1989**, *115*, 416–430. [CrossRef]
34. Wurbs, R.A. *Daily Water Availability Model for the Brazos River Basin and San Jacinto-Brazos Coastal Basin*; TR-513; Texas Water Resources Institute: College Station, TX, USA, 2019; 238p. Available online: https://twri.tamu.edu/publications/technical-reports/2019-technical-reports/tr-513/ (accessed on 1 October 2020).
35. Wurbs, R.A. *Daily Water Availability Model for the Trinity River Basin*; Texas Commission on Environmental Quality: Austin, TX, USA, 2019; 193p.
36. Wurbs, R.A. *Daily Water Availability Model for the Neches River Basin*; Texas Commission on Environmental Quality: Austin, TX, USA, 2020; 198p.
37. Texas Commission on Environmental Quality, Water Availability Modeling System. Available online: https://www.tceq.texas.gov/permitting/water_rights/wr_technical-resources/wam.html (accessed on 1 October 2020).
38. Hydrologic Engineering Center. *HEC-DSSVue HEC Data Storage System Visual Utility Engine, User's Manual*; Version 2, CPD-79; U.S. Army Corps of Engineers: Davis, CA, USA, 2009. Available online: https://www.hec.usace.army.mil/ (accessed on 1 October 2020).
39. Wurbs, R.A.; Lee, C.H. Salinity in water availability modeling. *J. Hydrol.* **2011**, *409*, 451–459. [CrossRef]
40. Wurbs, R.A.; Schnier, S.T.; Olmos, H.E. Short-term reservoir storage frequency relationships. *J. Water Resour. Plan. Manag.* **2012**, *138*, 597–605. [CrossRef]
41. United States Geological Survey (USGS) National Water Information System (NWIS). Available online: https://waterdata.usgs.gov/nwis (accessed on 1 October 2020).
42. Wurbs, R.A. Chapter 137 Institutional framework for water management. In *Handbook of Applied Hydrology*; Singh, V.P., Ed.; McGraw-Hill: New York, NY, USA, 2016.
43. Texas Commission on Environmental Quality. *Chapter 298 Environmental Flow Standards for Surface Water. Subchapter G: Brazos River and Its Associated Bay and Estuary System*; Texas Water Code: Austin, TX, USA, 2014.

Available online: https://www.tceq.texas.gov/assets/public/legal/rules/rules/pdflib/298g.pdf (accessed on 1 October 2020).
44. Brazos River Basin and Bay Expert Science Team. *Environmental Flow Regime Recommendations Report*; Final Submission to the Brazos River Basin and Bay Area Stakeholder Committee, Environmental Flows Advisory Group, and Texas Commission on Environmental Quality: Austin, TX, USA, March 2012. Available online: https://www.tceq.texas.gov/assets/public/permitting/watersupply/water_rights/eflows/brazos_bbest_complete_document.pdf (accessed on 1 October 2020).
45. Brazos River and Associated Bay and Estuary System Basin and Bay Stakeholders Committee. *Environmental Flow Standards and Strategies Recommendations Report*; Brazos River and Associated Bay and Estuary System Basin and Bay Stakeholders Committee: Austin, TX, USA, 2012. Available online: https://www.tceq.texas.gov/assets/public/permitting/watersupply/water_rights/eflows/brazos_bbasc_report_8_22_2012_bbasc.pdf (accessed on 1 October 2020).
46. Wurbs, R.A.; Ayala, R.A. Reservoir evaporation in Texas, USA. *J. Hydrol.* **2014**, *510*, 1–9. [CrossRef]
47. Muttiah, R.S.; Wurbs, R.A. Modeling the impacts of climate change on water supply reliabilities. *Water Int.* **2002**, *27*, 407–419. [CrossRef]
48. Wurbs, R.A.; Muttiah, R.S.; Felden, F. Incorporation of climate change in water availability modeling. *J. Hydrol. Eng.* **2005**, *10*, 375–385. [CrossRef]
49. Neilsen-Gammon, J.W.; Banner, J.L.; Cook, B.I.; Tremaine, D.M.; Wong, C.I.; Mace, R.E.; Gao, H.; Yang, Z.L.; Gonzales, M.F.; Hoffpauir, R.; et al. Unprecedented drought challenges for Texas water resources in a changing climate: What do researchers and stakeholders need to know? *J. Earth's Future* **2020**, *8*. [CrossRef]
50. Tarlock, A.D.; Corbridge, J.N.; Getches, D.A. *Water Resource Management: A Casebook in Law and Public Policy*, 6th ed.; Foundation Press: New York, NY, USA, 2009.
51. Getches, D.H. *Water Law in a Nutshell*, 4th ed.; Thomson/Reuters: St. Paul, MN, USA, 2009.

© 2020 by the author. Licensee MDPI, Basel, Switzerland. This article is an open access article distributed under the terms and conditions of the Creative Commons Attribution (CC BY) license (http://creativecommons.org/licenses/by/4.0/).

Article

The Role of Water Supply Development in the Earth System

Slobodan P. Simonovic * and Patrick A. Breach

Department of Civil and Environmental Engineering, The University of Western Ontario, London, ON N6A 3K7, Canada; pbreach@uwo.ca

* Correspondence: simonovic@uwo.ca

Received: 23 October 2020; Accepted: 24 November 2020; Published: 29 November 2020

Abstract: The ANEMI model is an integrated assessment model of global change that emphasizes the role of water resources. Securing water resources for the future is a key issue of global change and ties into global systems of population growth, climate change carbon cycle, hydrologic cycle, economy, energy production, land use and pollution generation. The focus of the presented work is on the development of global water supplies necessary to keep pace with a growing population and global economy. With the structure of the ANEMI model, a series of experiments are conducted in order to assess: (i) the current role of water supply in the global Earth system; (ii) the level of water stress that can be expected in the future; and (iii) what are the potential effects of water quality on global surface water supply and the distribution of water supply types. The results of model simulations show that surface water resources were sufficient to meet the water demand and water quality is not shown to be a significant factor for the development of surface water supplies. Due to globally aggregated scale, these impacts are averaged and likely understated.

Keywords: global change; integrated assessment modelling; system dynamics simulation; water resources management; water supply; climate change; earth system; feedback

---

1. Introduction

Human impacts on the environment at global scales are being realized through our ability to alter atmospheric concentrations of greenhouse gases and consequently global climate, creating the need to consider environmental problems and their interactions with the Earth as a highly integrated system. The Earth system is composed of biological, physical, chemical and human elements that form a network of feedbacks through their interconnections [1]. The concept of global change becomes increasingly important as the components of the Earth system such as population growth and migrations, economic productivity, climate, food production and hydrology are interlinked through dynamic non-linear feedback processes [2]. Within this system, changes in one component inevitably lead to changes in another. This is why global change research focusses on interactions between components of the Earth system as a whole, as opposed to only those of climate [1,3].

The main focus of the presented work is to answer the following questions through the implementation of the global model of the Earth system: (i) What level of water stress can be expected in the future?; (ii) Can alternative water supplies help to alleviate future water stress?; and (iii) What are the potential effects of water quality on global surface water supply and the distribution of water supply types?

1.1. Water in the Earth System

Water can be considered one of, if not the most, important drivers for human life as well as social and economic development [4]. Water resources provide for the most basic human needs of drinking and sanitation, while allowing for irrigated agriculture to take place and industrial activities such as thermal power generation, mining and manufacturing. Therefore, the use of, management and availability of water resources plays a crucial role in the progression of global changes in the Earth system as without it, societies cannot function.

A growing global population and its needs has put stress on water resources in many regions around the World. This problem will continue to grow as the population is projected (a) to increase 42% by the year 2100 to 10.9 billion people [5]; and (b) migrate–1 billion people may become climate refugees by 2050 [6]. The demand for water increases not only with the population but also with the consumption of water on a per capita basis. Alcamo et al. in Reference [7], show that countries with higher gross domestic product (GDP) per capita generally have higher water usage in the domestic sector and follow a type of S-curve, while in the industrial sector, water usage decreases exponentially to an equilibrium value. Therefore, as countries continue to develop economically the water usage patterns will change. By continuing with the current trends in global population, economics and technological change, water demand will continue to increase in most developing countries due increased domestic water usage as well as agricultural production. In developed countries domestic and industrial demands saturate and the expansion of irrigated land stagnates [8].

Water stress is often defined as the ratio of water withdrawals to the available water resources in a region. The hydrologic cycle along with changes made to it through anthropogenic means dictates the amount of water resources that are available for use. Although natural variability in weather patterns can determine if a region will experience wet or dry seasons, human influence on hydrologic cycles such as the construction and operation of dams and reservoirs, water diversions and water withdrawals redistribute the water availability in time and space. Climate change is expected to alter the spatial and temporal distribution of water resources on top of what is observed naturally and through direct human influence [9]. Increased global temperature through the greenhouse effect is expected to intensify the hydrologic cycle, leading to higher evapotranspiration rates, more frequent and heavier storms and faster flowing rivers, along with the potential for longer periods of drought. Because of this, there exists the potential for the availability of water resources to be changed for better or worse in different areas of the world [10].

Water resources may be available in a given point in time and space; however, the quality of that water can sometimes dictate whether or not it is available for a certain type of use. For example, according to a national report from the US Environmental Protection Agency almost half of rivers and streams across the US are categorized as with "poor biological condition" as a result of nutrient and sediment pollution. The condition of the rivers and streams are deemed unfit for fishing and recreational use [11]. In China, the situation is even worse with more than 70 percent of rivers and lakes being polluted and almost half may contain water unfit for human consumption or contact [12].

Degrading water quality over time has been shown to cause maintenance and treatment issues in drinking water treatment plants. There is evidence that increases in dissolved organic matter can lead to fouling and blocking membranes and filters, cause harmful disinfection by-products, facilitate biological re-growth in distribution systems and transport pesticides, pharmaceuticals and heavy metal into treatment systems [13]. There are a number of studies that highlight the relationship between the water quality and the water treatment, which can lead to water supplies inadequate for human consumption. Changes in water quality on a global scale could be a significant concern for our ability to maintain clean and sufficient water supplies.

Solutions to ensuring freshwater security vary from managing water demands and more accurately modelling water resource availability (surface and ground water), to technological solutions such as desalination and water reuse. Desalination involves the use of thermal evaporation or membrane separation technology to remove dissolved solids that are present in saline water sources. Both methods are highly energy intensive and can be costly when compared to traditional water supplies. Currently, there are approximately 16 thousand operational desalination plants around the World producing over 95 million m^3/day of desalinated water for human use [14]. The cost associated with producing this type of water supply is estimated to be between 0.45 to 2.51 $/m^3, which is still 2 to 3 times higher than conventional water supply but it is rapidly decreasing (approximately a factor of 10 since the 1960s [15].

Water reuse technologies involve the treatment of waste waters from a variety of different uses such as agricultural, municipal and industrial. The level of treatment necessary is dependent on the composition of waste waters being treated as well as the type of reuse that is under consideration. For non-potable reuse, wastewater is treated to a lower standard while potable uses require more advanced treatment methods capable of removing emerging pathogens, endocrine disrupting chemicals and pharmaceuticals [16]. Treatment options vary from simple low-energy solutions such as lagoons which allow wastewater to filter through media, to high-energy advanced treatment plants employing activated sludge treatment along with different levels of disinfection ranging from ultra-violet to membrane filtration.

1.2. Integrated Assessment Modelling

Water resources management in the context of global change involves many different disciplines ranging from climate science, economics, hydrology, biology, engineering, governance, agriculture and social sciences as outlined above. In order to address the problem of dealing with future water stress, these disciplines must be put together in a comprehensive framework. This will allow decision makers to explore policy options that consider the Earth system as a whole.

Assessment of various aspects of global change often requires the use of models from different domains and new tools and modelling paradigms to analyze complex interactions in the Earth system at a variety of spatial and temporal scales. The concept of integrated assessment (IA) has been defined as an interdisciplinary process of bringing together knowledge from different disciplines, adding value in contrast to a single disciplinary approach in order to provide information to decision and policy makers [17]. It is performed to bring about understanding of an issue regardless of the discipline.

Tol and Vellinga in Reference [18] describe the process of IA in a set of stages. The first stage involves structuring the problem that is to be assessed. The boundary of the problem must be defined in a way that encompasses all the important components of the problem, as well as components that may become important to the problem under different conditions or over time. Stage 2 involves the use of participatory and modelling methods for assessment to engage stakeholders that play a role in the problem at hand.

The integrated assessment modelling (IAM) approach involves the coupling of disciplinary models. There are many different methods that can be used to form a model for integrated assessment. Connections between disciplinary models can be made statically (output of one model is first obtained then given as input to another) or dynamically (both models running at the same time). The latter of which, is the only way that feedback loops can be created and studied. Dynamic connections can be made by using a computer program to facilitate the exchange of information while the models are running or both models can be combined into the same computer code [18].

1.3. System Dynamics Simulation for Integrated Assessment

The field of system dynamics focusses specifically on analyzing the dynamic nature of systems that are composed of feedback loops. Therefore, the use of system dynamics is ideal for the construction of integrated assessment models of global change. The system dynamics modelling process involves the use of causal loop diagramming to map out the feedback loops that are driving system behavior. This is effectively describing the boundary of the problem as well as the components that are responsible for reproducing it. System dynamics simulation builds from the conceptual models developed through systems thinking by adding structure to them. The addition of stocks or state variables and the flows that affect them, takes the system from a conceptual model to a mathematical model through stock and flow diagramming. Stock and flow diagrams illustrate the configuration of stocks and flows which is essentially a visual representation of a system of first order differential equations. Most, if not all,

IAMs can be represented in this way from a high level. Therefore, the system dynamics simulation approach is ideal for the construction of IAMs and provides a formalized way for creating feedback loops between disciplinary models of global change.

1.4. The Role of Water Supply Development in the Earth System

The ANEMI model [19,20] is an integrated assessment model of global change that emphasizes the role of water resources. The model is based on the principles of system dynamics simulation in order to analyze changes in the Earth system using feedback processes. Securing water resources for the future is a key issue of global change and ties into global systems of population growth, climate change carbon cycle, hydrologic cycle, economy, energy production, land use and pollution generation.

The main contribution of the presented work is the development of global water supplies necessary to keep pace with a growing population and global economy using an integrated feedback-based approach. With the structure of the ANEMI model, a series of experiments are conducted in order to assess: (i) the current role of water supply in the global Earth system; (ii) the level of water stress that can be expected in the future; and (iii) what are the potential effects of water quality on global surface water supply and the distribution of water supply types.

Evaluation of the model performance demonstrates that the model can reproduce historical trends related to global change within the Earth system. The experimental results show that investment in alternative water supplies on a global scale should be made in advance of conventional water supply depletion, as time delays may result in prolonged increases in global water stress. It was also found that the role of technological change was a greater factor for meeting future food production requirements than the effect of a changing climate. The impact of water quality degradation and the depletion of available water resource on water supply development, was found to be understated when studied on the global scale. It is recommended that the water supply development system developed in this work be extended to a finer spatial scale where the effects of water depletion and water quality degradation can be more thoroughly examined.

2. Overview of the ANEMI3 Model of Global Change

This chapter presents the ANEMI model, which is currently in version 3 [19,20], built upon the first two iterations of ANEMI [21,22]. The model shares the same system dynamics simulation paradigm that was used in the previous iterations of ANEMI, in that feedbacks and delays are used to drive system behavior. ANEMI3 is a type of integrated assessment model that describes the state of and interactions between model sub-systems that compose the Earth system. The main sub-systems or 'sectors' used are that of the climate system, carbon, nutrient and hydrologic cycles, population dynamics, land use, food production, sea level rise, energy production, global economy, persistent pollution, water demand and water supply development.

Each individual sector in the model describes the relevant feedbacks that drive the state variables in the sector. Connections between sectors form intersectoral feedbacks responsible for the functioning of the Earth system. It is the intersectoral feedbacks that allow us to represent feedbacks that drive global changes in the Earth system. Feedbacks driving global change are now evident, while is expected that negative feedbacks acting on population and economic growth may be more evident in the future. From a system dynamics perspective, effective policymaking should be based on addressing the feedback structure of a system, not only on modifying the system parameters. This viewpoint is what makes the ANEMI3 model unique and useful since in the current time global modelling is becoming progressively more complex [23].

The boundary of the model is defined by the problem that is being explored. In this case, we are modelling the role of water resources in various aspects of global change. Therefore, the spatial scale of the model is mainly one that is global. In some sectors, the stocks are disaggregated to capture material flows on a sub-global scale but not at a level that is location specific. This spatial scale limits the level of detail that can be used to describe the flows that act to change the model stocks, however it allows us to effectively analyze feedbacks between water resources and other model sectors on a global scale.

The highly endogenous structure and coupling of sub-systems in the ANEMI3 model are part of its novelty in the realm of integrated assessment modelling. Because of this, feedback processes are responsible for the behavior that is exhibited in model runs. The model sectors that comprise the ANEMI3 model are that of the climate system, carbon, nutrient and hydrologic cycles, population dynamics, land use, food production, sea level rise, energy production, global economy, persistent pollution, water demand and water supply development as shown in Figure 1. The model includes over 2000 variables and 700 equations. Presentation in Figure 1 is focused on illustrating the high-level model sectoral structure and relationships. Feedback loops between sectors or intersectoral feedback loops are responsible for global change in this Earth system. Intersectoral feedbacks in the ANEMI3 model allow for the representation of various aspects of global change. In the Figure 1 diagram alone there is a total of 89 possible intersectoral feedback loops. The size of the feedback loops range from 2 to 9 sectors included out of the 10 that are shown. An example is that of a growing global economy, which drives energy production and industrial growth, thereby resulting in more greenhouse gas emissions and climate change. This in turn results in negative feedbacks on economic growth through climate damages, which can represent economic damages because of land and structures lost to coastal flooding, for example. Creating a causal loop diagram from these connections between model sectors allows us to view the feedbacks that are created by combining model sectors in this way.

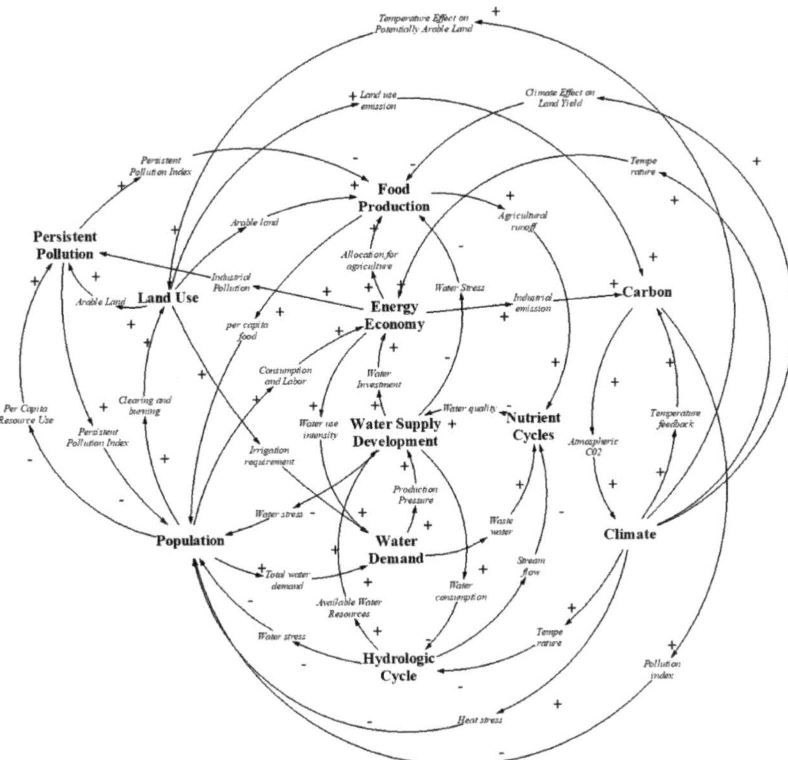

Figure 1. High-level feedback structure of the ANEMI3 model illustrated as a causal loop diagram (+ signs along causal relationships indicate change of connected variables in the same direction; − sign indicates change in opposite direction).

The main difference between ANEMI1, ANEMI2 and ANEMI3 is addition of intersectoral feedback loops used to (a) analyze water supply development within the Earth system, (b) include of water quality degradation and its impact on the development of surface water supplies and (c) assess of global scale feedback related to water supply development. These main modifications are introduced to represent the dynamics of global change at the global scale with an emphasis on the development of water supplies.

The ANEMI model is developed using Vensim (https://vensim.com/ last accessed 20 November 2020) system dynamics simulation software. The current model code is archived using Zenodo (https:doi.org/10.5281/zenodo.4025424) and details on how to run the model, modify inputs and view the outputs in graphical or tabular formats are provided in the repository and discussed in Reference [19]. Work presented in this paper is building on the work and data from previous versions of the model [15,16,19].

2.1. Integrated Assessment of Global Water Resources

The new water supply sector in ANEMI3 was developed by incorporating water supply as a new production sector within the newly added energy-economy sector [19,20,23]. This has been achieved by adding capital stocks to produce water supply in the form of surface, ground, wastewater reclamation and desalination water sources.

As available water resources become depleted, the water supply is reduced for the same input intensity. This means that more effort is required to produce the same rate of water supply, which also makes a given type of water supply that is depleted more expensive. For example, when the groundwater elevation decreases from over abstraction, more pumping energy is required to extract the same amount of water resource. The effect of saturation is also included in this relationship, assuming the best or most cost-effective sites are used first for water supply infrastructures. An example of which could include the construction of additional reservoirs, source water intakes, of groundwater wells in areas that are less suitable or cost effective than those that were previously constructed.

The dotted causal link from water price to the capital order rate in Figure 2 indicates a connection that is neither positive nor negative. Instead, this link is used to determine the amount of investment that is made in the capital stocks of the different supply types (surface, ground, wastewater reclamation and desalination water sources). Inputs from the nutrient cycle, hydrologic cycle and water demand sectors are used to define the water price, water stress and water resource ratio variables respectively in the water supply development sector.

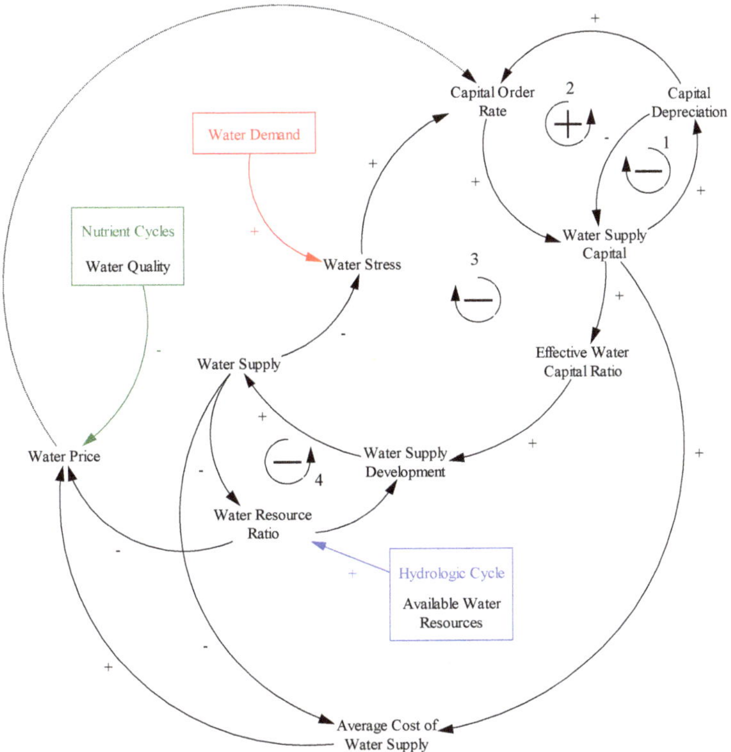

Figure 2. Causal loop diagram of the ANEMI3 water supply development sector. The dotted arrow from water price to water supply indicates a causality that is neither positive nor negative. Different colors identify inputs coming from different model sectors. Clockwise arrow with – sign designates negative feedback loop and counter clockwise arrow with + sign designates positive feedback loop.

2.2. Mathematical Formulation of Water Supply Development Sector

Water resources, R_i are used in the production of water supplies, where the subscript i, denotes the type of water supplies for which the water resources are being used.

$$R_{sw} = S_r \times TRF - URW \times WPF \left[\mathrm{km}^3/\mathrm{y}\right] \tag{1}$$

$$R_{gw} = Q_{perc} - Q_{discharge} \left[\mathrm{km}^3/\mathrm{y}\right] \tag{2}$$

$$R_{ww} = TDW + TIW \left[\mathrm{km}^3/\mathrm{y}\right] \tag{3}$$

$$R_{ds} = Oceans \left[\mathrm{km}^3\right], \tag{4}$$

where R_{sw} = Surface water resources $\left[\mathrm{km}^3/\mathrm{y}\right]$; R_{gw} = Groundwater resources $\left[\mathrm{km}^3/\mathrm{y}\right]$; R_{ww} = Wastewater resources $\left[\mathrm{km}^3/\mathrm{y}\right]$; R_{ds} = Desalination water resources; S_r = Stable and reusable runoff fraction; TRF = Total renewable flow $\left[\mathrm{km}^3/\mathrm{y}\right]$; WPF = Wastewater pollution factor; Q_{perc} = Percolation to groundwater $\left[\mathrm{km}^3/\mathrm{y}\right]$; $Q_{discahrge}$ = Groundwater discharge $\left[\mathrm{km}^3/\mathrm{y}\right]$; TDW = Treated domestic wastewater $\left[\mathrm{km}^3/\mathrm{y}\right]$; TIW = Treated industrial wastewater $\left[\mathrm{km}^3/\mathrm{y}\right]$; URW = Untreated Returnable Waters $\left[\mathrm{km}^3/\mathrm{y}\right]$.

The amount of water resources available for the development of water supplies is dependent on the hydrologic cycle, water demand and water quality sectors of the model. In the case of surface water, the stable and reusable portion of runoff is taken from the total renewable streamflow and is adjusted for untreated wastewater discharge. The adjustment for wastewater discharge is based on [24] which estimates that for every cubic meter of contaminated wastewater discharged into water bodies and streams, makes unsuitable 8–10 cubic meters of fresh water. The difference in groundwater percolation and discharge is used for the consideration of groundwater resources as this refers to renewable groundwater. Only renewable groundwater resources are considered for the global scale. The inclusion of non-renewable or fossil groundwater resources should be considered at the regional scale. For the potential reuse of wastewater, industrial and domestic wastewaters are considered. Although the reuse of wastewater is highly dependent on the type of wastewater and the use for which it is being treated, it is considered here as a supplementary type of water supply in the case of groundwater and surface water depletion. Water resources used for desalination are considered primarily from the ocean stock in the hydrologic cycle. This results in a virtually limitless supply; however, it is very energy intensive resulting in a high effective input intensity thereby limiting production.

The concept of resource depletion in energy production is also applicable to water supply development. For example, in the case of surface water and groundwater resources, depleted water resources will mean less suitable locations for water extraction and treatment plants. This might mean that source waters could be further from where the water is being used, thus increasing distribution costs. Pumping costs could also be increased by using deeper aquifers or surface water supplies that have a greater difference in elevation from their point of use. Water resource depletion factors into the water supply development process in much the same way as energy production, however there is one key difference. The depletion effect for energy production is based on the ratio of current energy resources remaining to the initial amount. In contrast, water resources are renewable to varying degrees. Therefore, simply taking the ratio of the available water resources to the initial water resources is insufficient. Here, the ratio of available water resources to the current production level is used. In order to accomplish this structure, water production was changed to a stock variable to avoid creating an indeterminate system (introduction of a new negative feedback by making water production a function of itself).

$$WS_i = \int WS_{i,0}\left(\alpha_{w_i}\left(\frac{WS_i}{AW_i}\right)^{\rho_{wi}} + \left(1-\alpha_{w_i}\right)EWII_i^{\rho_{wi}}\right)^{\frac{1}{\rho_{w_i}}} \times dt\left[\text{km}^3/\text{y}\right], \tag{5}$$

where WS_i = Water supply from water resource i $\left[\text{km}^3/\text{y}\right]$; $WS_{i,0}$ = Initial water production $\left[\text{km}^3/\text{y}\right]$; AW_i = Available water resource remaining $\left[\text{km}^3/\text{y}\right]$; $EWII_i$ = Effective water input intensity; α_{wi} = Water resource share; ρ_{wi} = Resource substitution coefficient.

In the case of surface water, the available water resources are a rate (runoff minus water quality depletion effects) rather than a stock that can be depleted over time. If production equals this rate, then there is no more surface water that can be utilized at this time step. For wastewater reuse if the rate of reuse is equal to that of the amount of treated wastewater, then no more wastewater can be reused unless wastewater treatment percentage increases.

In the energy capital sub-system of the energy-economy sector, the desired energy capital for each source is determined by the perceived return on investment and the production pressure defined as the ratio of the energy order rate or demand to energy production for each source [19,20]. In the case of water supply, the term for perceived return on investment is removed, thereby making the primary drive for new water supply capital based on production pressure, which resembles the definition of water stress (withdrawal or demand to availability ratio). This value is multiplied by the current water capital stocks to obtain the desired water capital stocks,

$$DKW_i = KW_i \times \frac{W_{di}}{WS_i}[\$], \tag{6}$$

where DKW_i. = Desired water capital for water source $i \left[km^3/y\right]$; W_{d_i} = Demand for water supply i $\left[km^3/y\right]$; WS_i = Water supply from water source $i \left[km^3/y\right]$.

Where i denotes the type of water supply for which desired water capital is being determined. In order to obtain the demand for water supply from each source, Wood's algorithm [21] is used to allocate the total water demand (sum of domestic, industrial and agricultural water demand) to each supplier. The geometric illustration of Wood's algorithm is shown in Figure 3, where each rectangle represents a different supplier (blue-surface, orange-ground, green-wastewater reclamation and red-desalination water supplies). The area of each rectangle represents the capacity for a given supplier to fulfil the demand for a product, while the position and width of each rectangle is based on the "attractiveness" value and "width" parameters respectively. Here, the inverse water supply price is used to represent the attractiveness value and the area of each rectangle would be the water supply capacity for a given supply type. The total water demand is allocated to each supplier by the black line in Figure 3 which moves from right to left until the area to the right of the line fulfils the demand. The area of each rectangle that lies on the right of the black line represents the level of demand satisfied by each supplier, therefore a water supply type with a high price would be placed farther to the left on the attractiveness scale and would receive less of the total water demand.

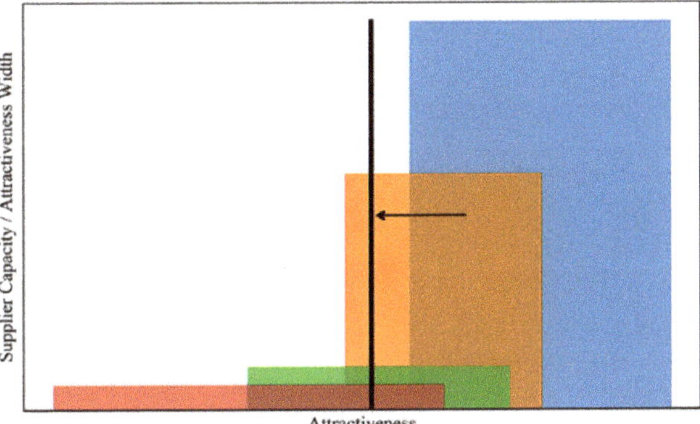

Figure 3. Illustration of Wood's algorithm.

The inverse water supply price was chosen as the main driver for changes in supplier attractiveness as this will vary with technological improvements, depletion, saturation and water quality in the case of surface water supply. This formulation encapsulates the effects of global changes in technology, water resource availability and water quality on the allocation of capital investments in different types of water supply. The width factor determines how this allocation is distributed to suppliers which are not necessarily the cheapest option. For example, on the global scale, although the use of surface water supplies is likely the most cost-effective option in many regions, groundwater, water reuse and desalination supplies are all being used simultaneously.

The concept of endogenous technological change applied to energy production [19,20] has analogies to water supply development. In the case of surface water and groundwater supplies, it is assumed that pumping, distribution and treatment technologies will remain largely the same but will show some improvement over time. However, alternative water supplies such as wastewater reuse and desalination are likely to see vast improvements in the near future. Factoring technological change into the water supply development process is what will help make alternative water supplies more feasible in the future, along with depletion and saturation of conventional water supplies.

A unique attribute of water resources when considering water supply development is water quality. Degraded water quality can impact the functioning of water treatment facilities as well as maintenance costs and the necessary configuration of unit processes [22]. This may also influence the ability to secure adequate source waters for extraction of water resources in the future as a result of pollution and climate change. This could negatively impact production of conventional water supplies by increasing the cost of implementing new capital as well as variable inputs needed for treatment and distribution including energy, chemicals and labor.

In ANEMI3, nutrient concentrations in surface waters are used as an indicator of water quality on a global scale [19,20,22]. Wastewater and agricultural inputs are used as the main contributors to water quality degradation and changes in the levels of nutrients in the form of total nitrogen and phosphorus are used as indicators of water quality from the nutrient cycle sector of the model. The ratio of current to initial nutrient concentrations for surface water resources is used as a multiplier on the water supply price,

$$P_{w_{sw}} = PP_{w_{sw}} \times \left(\frac{NCE}{NCE_0}\right)^{\gamma_w} \left[\$/\text{km}^3\right], \tag{7}$$

where $P_{w_{sw}}$ = Water supply price for surface water $\left[\$/\text{km}^3\right]$; $PP_{w_{sw}}$ = Producer price for surface water $\left[\$/\text{km}^3\right]$; NCE = Nutrient concentration effect $\left[(nN \times nP)/\left(\text{km}^3/\text{y}\right)^2\right]$; NCE_0 = Initial nutrient concentration effect $\left[(nN \times nP)/\left(\text{km}^3/\text{y}\right)^2\right]$; γ_w = Influence of water quality on surface water supply price.

The nutrient concentration effect takes into consideration the concentration of both total nitrogen and phosphorus,

$$NCE = \frac{N_{N_{River}} \times N_{P_{River}}}{SF^2} \left[(nN \times nP)/\left(\text{km}^3/\text{y}\right)^2\right], \tag{8}$$

where $N_{N_{River}}$ = Nitrogen content of river stock [nN]; $N_{P_{River}}$ = Phosphorus content of river stock [nP]; SF = Streamflow $\left[\text{km}^3/\text{y}\right]$.

2.3. Integrating Water Supply Development Sector into ANEMI3 Structure

In order to include water supply development as an additional component within the energy-economy sector, key connections needed to be made with the energy-economy sector of the model. Those connections are detailed in Figure 4. Establishing these connections effectively closes several feedback loops for water supply development to fit into this sector. Water supply development is treated as an additional horizontal disaggregation of the global capital stock alongside the energy sector.

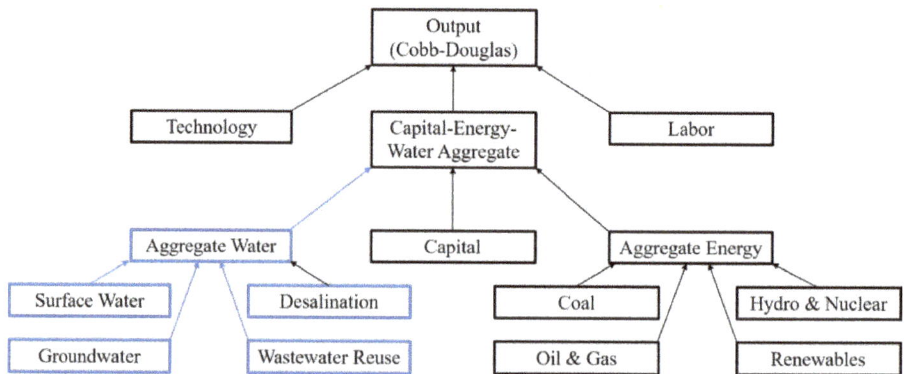

Figure 4. Production structure of water supply within the energy-economy-water sector of the ANEMI3.

To accomplish this production structure, water production, capital, technological change and pricing structures were replicated from that of the energy economy sector. Capital stocks were created to represent water supply infrastructures for surface water, groundwater, wastewater reuse and desalination. The level of capital for each source refers to any infrastructure that relates to the global capacity of the system to provide water supply. This includes reservoirs, pumping systems, treatment systems and distribution networks. Economic output in the energy-economy sector is distributed amongst energy and water production, investment and consumption. The inclusion of water supply development adds an additional consumer of economic output (Figure 5).

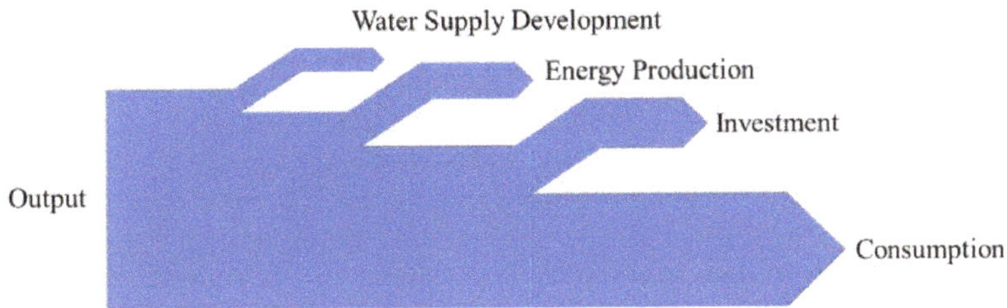

Figure 5. Goods allocation in the energy-water-economy sector of theANEMI3.

3. Model Experiments

To assess future levels of water stress and the role of alternative water supplies and water quality, three experiments are used with ANEMI3 model. In the first, different formulations of water stress are compared to examine the driving factors of water stress on a global scale. The second experiment focusses on development pathways of alternative water supplies including water reclamation and reuse and desalination. Different development pathways are examined to estimate whether it is possible that sufficient supplies can be developed to alleviate global water stress. The final experiment is used to examine the potential effect of water quality on surface water supply. Here an indicator of global water quality is used to alter the production of surface water supplies, assuming that significantly lower water quality source waters are more costly to make available to the population. Each of the three experiments is discussed in detail below.

3.1. Experiment 1—Examination of Future Global Water Stress

Thresholds of water stress have been defined by Reference [23]. Low, moderate, medium-high and high levels of water stress corresponds to values of less than 0.1, 0.1 to 0.2, 0.2 to 0.4 and greater than 0.4 respectively, where water stress (WTA) is defined as the ratio of surface water withdrawals (SWW) to availability (ASW),

$$WTA = \frac{SWW}{ASW}. \tag{9}$$

In the ANEMI3 model, water stress can be calculated using different formulations. Water pollution and green water dilution effects (WTA_{poll} and $WTA_{poll+gw}$ can be applied to the WTA ratio in order to gain a more conservative measure of water stress [24].

$$WTA_{pollution} = \frac{SWW + URW \times WDF}{TotalRenewableFlow} \tag{10}$$

$$WTA_{pollution+gw} = \frac{SWW + URW \times WDF + GWR}{TotalRenewableFlow}, \tag{11}$$

where URW = Untreated returnable water $\left[\mathrm{km}^3/\mathrm{y}\right]$; WPF = Water pollution factor; GWR = Green water requirement for crops and pasture $\left[\mathrm{km}^3/\mathrm{y}\right]$.

In this work, an additional representation is used based on the ratio of total water supply to the amount of available conventional water resources of surface water (R_{sw}) and groundwater (R_{gw}).

$$WTA_{watersupply} = \frac{\sum WS_i}{R_{sw} + R_{gw}}. \tag{12}$$

The total amount of water supply includes both, conventional and alternative water resources, allowing for increased alternative water resources to reduce water stress.

3.2. Experiment 2—The Role of Alternative Water Supplies

Growing populations and industrial output will increase the demand for water in the domestic, industrial and agricultural sectors, thereby increasing the pressure on freshwater resources. It is expected that these resources will become increasingly stressed over time, such that the ratio of demand to available water resources will increase. To overcome water stress, alternative supplies in addition to conventional surface water and groundwater will be needed, such as desalinated water and the wastewater reuse. The ability to analyze the distribution of water supplies through time will provide insight as to when the water resources become stressed and to what degree alternative water supplies will be needed in the future.

Alternative water supplies are represented in ANEMI3 in the same way as conventional water supplies including surface water and groundwater. However, the supply price starts at a higher value initially and is gradually reduced through improvements to the technology over time. The cost of producing alternative water supplies has decreased historically and is expected to decrease further. The rate at which technology improves in a complex system cannot be simply calculated, therefore the role of alternative water supplies in reducing future levels of water stress is examined through a Monte Carlo sensitivity analysis. The parameters used to specify technological change rates for alternative water resources is expressed using a probability distribution and the ANEMI3 model is then simulated 200 times to evaluate a range of pathways for alternative water supply development.

3.3. Experiment 3—Water Quality Effects on Surface Water Supplies

Water quality in ANEMI3 is represented by the changing concentrations of nutrient levels in surface waters. It acts as a multiplier that increases the supply price of surface water resources through hypothesized cost of increased treatment. This hypothesis is supported by the studies mentioned previously [22] but the extent of this effect is unknown and has never been looked at on a global scale. In addition to increased nutrients, wastewater inputs also render a portion of water resources unusable for the purpose of water supply, thereby contributing directly to water stress. If water quality becomes severely degraded in the future on a global scale, costs to produce water supplies could increase if technology does not progress fast enough to address potential treatment issues. Because of this, it is hypothesized that alternative water supplies may become more attractive and play a larger role in the future.

In ANEMI3, nutrient concentrations in surface waters are used as an indicator of water quality on a global scale. Wastewater and agricultural inputs are used as the main contributors to water quality degradation and changes in the levels of nutrients in the form of total nitrogen and phosphorus are used as indicators of water quality from the nutrient cycle sector of the model. The ratio of current to initial nutrient concentrations for surface water resources is used as a multiplier on the water supply price,

$$P_{w_{sw}} = PP_{w_{sw}} \times \left(\frac{NCE}{NCE_0}\right)^{\gamma_w} \left[\$/\mathrm{km}^3\right], \tag{13}$$

where $P_{w_{sw}}$ = Water supply price for surface water $\left[\$/\mathrm{km}^3\right]$; $PP_{w_{sw}}$ = Producer price for surface water $\left[\$/\mathrm{km}^3\right]$; NCE = Nutrient concentration effect $\left[\frac{nN \cdot nP}{(\mathrm{km}^3/\mathrm{y})^2}\right]$; NCE_0 = Initial nutrient concentration effect

$\left[\frac{nN \cdot nP}{(\mathrm{km}^3/\mathrm{y})^2}\right]$; γ_w = Influence of water quality on surface water supply price. The nutrient concentration effect takes into consideration the concentration of both total nitrogen and phosphorus,

$$NCE = \frac{N_{N_{River}} \times N_{P_{River}}}{SF^2}\left[(nN \times nP)/\left(\mathrm{km}^3/\mathrm{y}\right)^2\right], \tag{14}$$

where $N_{N_{River}}$ = Nitrogen content of river stock [nN]; $N_{P_{River}}$ = Phosphorus content of river stock [nP]; SF = Streamflow $\left[\mathrm{km}^3/\mathrm{y}\right]$.

The effect of water quality on water supply development is examined by comparing development pathways under different levels of nutrient inputs to surface waters via wastewater. Wastewater treatment rates are set constant and compared to the baseline wastewater treatment levels.

4. Results

This section presents the results of ANEMI3 model simulations performed to address the three research questions.

4.1. Experiment 1

The projected water stress values using the formulations mentioned above are shown in Figure 6. When the effects of pollution and green water dilution are included, water stress values are much higher. Using only the WTA ratio (Equation (9)), water stress values start initially at a value of 0.21 and rise up to 0.24, which is on the low end of the medium-high water stress category. In contrast, when pollution and green water effects are considered (Equation (11)), the starting values range between 0.32 to 0.35.

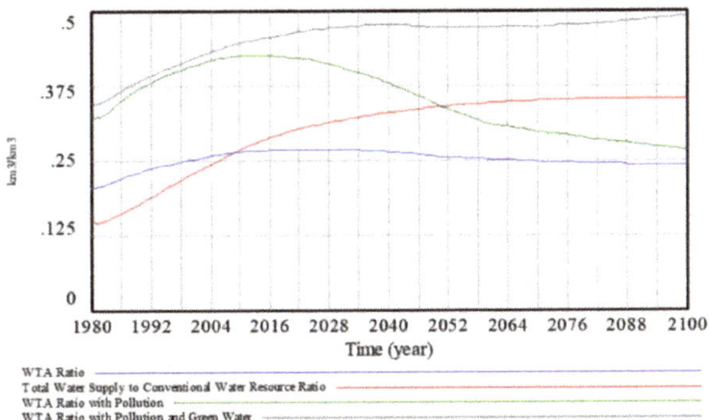

Figure 6. ANEMI3 simulated levels of water stress using the withdrawal to availability ratio and alternate formulations.

As the simulation progresses, water stress with only pollution effects considered (Equation (10)) on top of the WTA reaches a peak in the year 2010 and declines afterwards. This is because in this case the pollution effects are represented only through wastewater inputs, which decrease as domestic and industrial water demands decrease in the model due to reduced water intensities with greater global economic output. When water pollution in the form of agricultural runoff or green water is included, water stress values continue to rise to a value of 0.5 by the end of the simulation. This indicates severe levels of water stress. Using the ratio of water supply to available water resource levels as an indicator of water stress results in a starting value of 0.15 which follows S-shaped growth to 0.35. This indicates a shift from low levels of water stress to the high end of the medium-high water stress category.

4.2. Experiment 2

The development of water supplies for surface water, groundwater, wastewater reuse and desalination under the ANEMI3 baseline scenario are shown in Figure 7. Surface water supplies on a global scale have made up the largest fraction of water supply along with groundwater resources. They are the least costly to find and extract and there is much more capital currently invested in these supply types. However, in places where rivers or streams are not present, groundwater may be a less costly option, especially if the quality of the surface water is poor. Surface water supplies start at an initial value of 1504 km^3/year and climb to a maximum of 4422 km^3/year. Groundwater supplies increase at a much slower rate from 877 km^3/year to 1439 km^3/year. Both wastewater reuse and desalination supplies increase at a rate that is much faster than surface and groundwater, however the amounts of which are also much smaller initially, with wastewater reuse and desalination reaching 292 and 87 km^3/year by the end of the century, respectively.

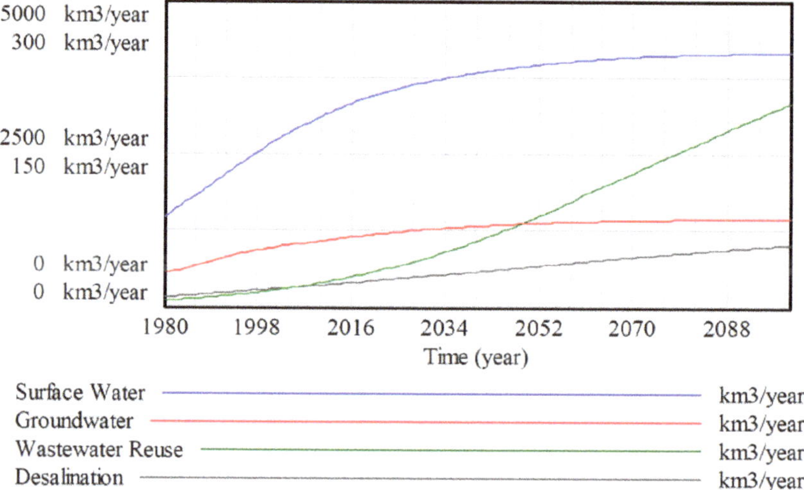

Figure 7. Development of water supplies in the ANEMI3 model. The upper scale labels are used for surface water and groundwater supply while the lower labels are for wastewater reuse and desalination.

Surface water supplies are the dominant source of water supply globally for the ANEMI3 baseline run. This is because the supply is relatively inexpensive and abundant, compared to the other water sources on a global scale. However, this is not always the case on a regional level. There are many areas of the world where either surface or groundwater resources are currently depleted or unavailable in time and space, thus prompting the use of alternative water resources, such as desalination and wastewater reuse.

4.3. Experiment 3

The input of nitrogen to surface waters is increasing throughout the baseline simulation starting at an initial rate of 3.1 trillion moles or 4.3 Mt per year to a rate of 7.6 trillion moles or 10.5 Mt per year (Figure 8). Input of phosphorus to surface waters on the other hand, increases from 451 billion moles or 13.5 Mt per year to a peak value of 681 billion moles or 20.4 Mt per year in the year 2025. After this point phosphorus input decreases significantly, down to 126 billion moles per year.

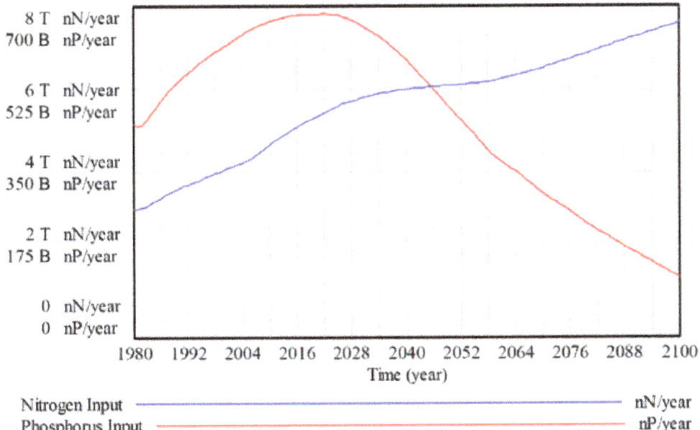

Figure 8. Total nitrogen and phosphorus input to surface water under the ANEMI3 baseline scenario. Left axis represents number of moles of nitrogen and phosphorus inputs to surface water per year.

The explanation for the difference in the pattern of nitrogen and phosphorus inputs lies in their respective amounts in different sources. For nitrogen, on a global scale, agriculture is the main anthropogenic source of nutrients to surface waters, while domestic and industrial wastewaters are the main source of phosphorus. Phosphorus input decreases after the year 2025 due to increasing levels of wastewater treatment on a global scale, which reduces the input significantly. The levels of treated and untreated wastewater are shown in Figure 9. Initially, the amount of untreated wastewater is greater than treated on a global scale in 1980. Under the ANEMI3 baseline scenario, wastewater treatment increases from the initial rate of 160 km^3/year and surpasses that of the untreated percentages in 2010. After this point, treatment rate increases further to approximately 550 km^3/year.

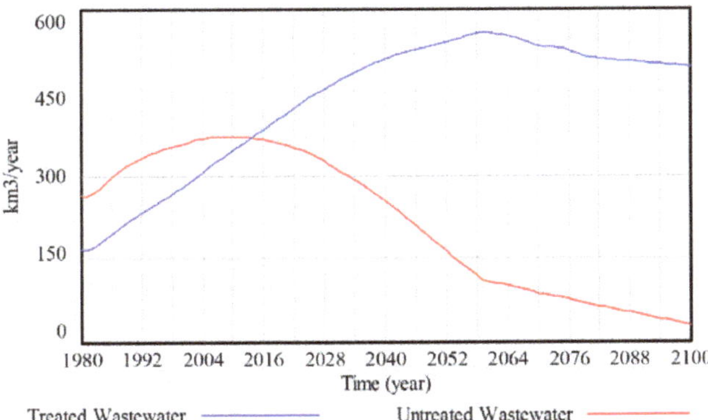

Figure 9. Treated and untreated wastewater inputs to the nutrient cycles over time.

Nutrient inputs act as an additional rate that affects the surface water stock in the nutrient cycle model. Combining this with the stock of surface water in the hydrologic cycle model allows for the concentrations of nutrients in surface water on a global scale to be examined, as shown in Figure 10. The concentration considers changes in hydrologic cycle. The patterns are almost the same because the global amount of streamflow does not change very much due to climate change increase and surface water consumption having a balancing effect in the ANEMI3 baseline scenario.

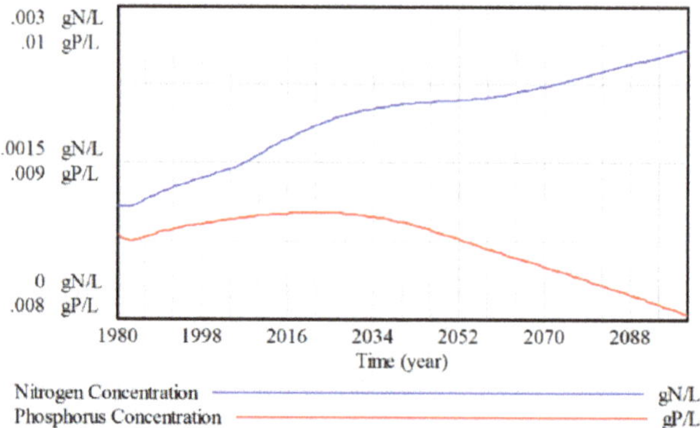

Figure 10. Surface water nutrient concentrations of nitrogen and phosphorus.

Nutrient concentrations are higher when constant wastewater treatment is implemented, rather than exogenous increase in the ANEMI3 baseline scenario. Nutrient concentrations are used as an indicator for water quality in the production of surface water supplies, whereby higher concentrations act as a multiplier to the surface water production costs. The effect of constant wastewater treatment on water supply development is shown in Figure 11. Under this scenario, the establishment of surface water supplies is only slightly affected by the change in surface water quality on a global scale (Figure 11a). Under the ANEMI3 baseline parameterization scheme, water quality does not appear to play a significant role in the establishment of surface water supplies, even if wastewater treatment levels are held at constant 1980 values for the entire simulation. Both wastewater reuse and desalination supplies show major increases from 1980 to the year 2100. Wastewater reuse increases from 10 to 280 km^3/year, while desalination increases from 10 to 75 m^3/year, although the absolute numbers are small in comparison to conventional water supplies. With reduced wastewater treatment rates there is a major difference in the level of wastewater reuse, as there is less available wastewater resource to be used (Figure 11b). Due to scarce wastewater for reuse there is a drop from 274 km^3/year to 143 km^3/year by the year 2100.

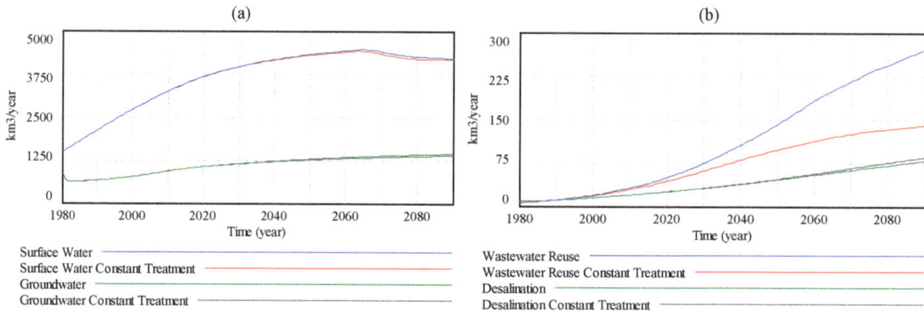

Figure 11. Development of water supplies under the baseline and constant wastewater treatment scenarios for (**a**) conventional water supplies and (**b**) alternative water supplies.

5. Discussion

The paper explores the utility of adding the feedback-driven, economically based water supply development sector in ANEMI3 global change model. Capital stocks for each type of water supply grow over time with investment, which is made based on the inverse supply prices and allocated using Wood's algorithm. Endogenous technological change is also incorporated for the desalination and wastewater reuse technologies, as well as the effects of depletion and diminishing water quality of conventional supplies.

In the ANEMI3 baseline scenario, water stress values are decreasing due to technological change and investments in water supply capital over time. The ANEMI3 baseline simulation for the development of water supplies shows that surface water resources are dominating the share of water supply during the entire simulation period from the year 1980 to 2100. This is because surface water resources are by far the least expensive option for water supply in the ANEMI3 baseline scenario. When only the global scale is considered, there is enough stable and renewable surface water resources to satisfy the demand of a growing population by the year 2100.

The potential for water quality impacts on the development of surface water supplies is assessed. Nutrient concentrations in surface water resources is calculated using the global cycles of water, nitrogen and phosphorus. The difference in sources of nitrogen and phosphorus inputs to the nutrient cycles, result in different long-term behaviors in their respective surface water concentrations.

6. Conclusions and Future Work

The ANEMI3 model structure is novel in that global water supply is able to evolve endogenously and allows for the development of conventional and alternative water supplies, while including effects of water quality on surface water resources. The development of water supply infrastructure is assessed from an economic perspective.

6.1. Main Conclusions

Under the current parameterization scheme, water quality is not shown to be a significant factor for the development of surface water supplies. When wastewater treatment rates are fixed at their initial values, surface water nutrient concentrations increase but not enough to show large impacts on surface water production.

Using increased nutrient concentrations as an indicator for water quality provides a way to represent the impact of different sources of water pollution but on a globally aggregated scale these impacts are averaged and likely understated. The reduced wastewater treatment scenario did however influence wastewater reuse. The lower quantity of treated wastewater available for reuse resulted in a greater saturation effect on the development of water supplies from wastewater reuse, thereby reducing its potential to develop as an alternative water resource.

6.2. Future Work

There are some limitations in presenting dynamics of the water supply development sector incorporated into ANEMI3 model on the global scale in the baseline ANEMI3 scenario. This is because surface water resources were enough to sustain the water demand when the available water resources consider the entire amount on Earth. This was also true for water quality, as it is averaged across the globe as well.

If the water supply development model is regionalized or adapted for use in a grid-based model, the effects of resource depletion and water quality effects on surface water supply could be explored in more detail. This is selected to be the major direction for future work. In doing this, location specific details with regards to water supply development could be considered, such as distribution costs for areas that are further away from coastlines in the case of desalination or the depth of regional aquifers for groundwater extraction costs.

Author Contributions: Conceptualization, S.P.S. and P.A.B.; software development and formal analysis, P.A.B.; writing—original draft preparation, S.P.S. and P.A.B.; writing—review and editing, S.P.S.; supervision, project administration and funding acquisition, S.P.S. Both authors have read and agreed to the published version of the manuscript.

Funding: This research has been funded by the Discovery Grant to the first author by the Natural Sciences and Engineering Research Council of Canada.

Acknowledgments: The authors are thankful for the constructive suggestions provided by the reviewers of the initial draft.

Conflicts of Interest: The authors declare no conflict of interest.

References

1. Steffen, W.; Sanderson, R.A.; Tyson, P.D.; Jäger, J.; Matson, P.A.; Moore III, B.; Oldfield, F.; Richardson, K.; Schellnhuber, H.J.; Turner, B.L.; et al. *Global Change and the Earth System: A Planet under Pressure*; Springer: Berlin/Heidelberg, Germany, 2004.
2. Davies, E.G. Modelling Feedback in the Society-Biosphere-Climate System. Ph.D. Thesis, University of Western, London, ON, Canada, 2007; p. 359. Available online: http://citeseerx.ist.psu.edu/viewdoc/download?doi=10.1.1.615.378&rep=rep1&type=pdf (accessed on 19 October 2020).
3. Cox, P.; Nakicenovic, N. Assessing and Simulating the Altered Functioning of the Earth System in the Anthropocene. In *Earth System Analysis for Sustainability*; Schellnhuber, H., Crutzen, P., Clark, W., Claussen, M., Held, H., Eds.; MIT Press: Cambridge, MA, USA, 2004; pp. 293–3124.
4. Rogers, P.; Bhatia, R.; Huber, A. *Water As a Social and Economic Good: How to Put the Principle into Practice*; Global Water Partnership Technical Advisory Committee, Global Water Partnership/Swedish International Development Cooperation Agency: Stockholm, Sweden, 1988.
5. United Nations. World Population Prospects 2019: Highlights. *Dep. Econ. Soc. Aff.* **2019**, 1–46. Available online: https://population.un.org/wpp/Publications/Files/WPP2019_Highlights.pdf (accessed on 8 October 2020).
6. Trimarchi, M.; Gleim, S. 1 Billion People May Become Climate Refugees by 2050. HowStuffWorks. 22 September 2020. Available online: https://science.howstuffworks.com/environmental/green-science/climate-refugee.htm (accessed on 8 October 2020).
7. Alcamo, J.; Döll, P.; Henrichs, T.; Kaspar, F.; Lehner, B.; Rösch, T.; Siebert, S. Development and testing of the WaterGAP 2 global model of water use and availability. *Hydrol. Sci. J.* **2003**, *48*, 317–337. [CrossRef]
8. Alcamo, J.; Döll, P.; Henrichs, T.; Kaspar, F.; Lehner, B.; Rösch, T.; Siebert, S. Global estimates of water withdrawals and availability under current and future "business-as-usual" conditions. *Hydrol. Sci. J.* **2003**, *48*, 339–348. [CrossRef]
9. Simonović, S.P. *Managing Water Resources: Methods and Tools for a Systems Approach*; UNESCO, Paris and Earthscan James & James: London, UK, 2012.
10. Schlosser, C.A.; Strzepek, K.; Gao, X.; Fant, C.; Blanc, É.; Paltsev, S.; Jacoby, H.; Reilly, J.; Gueneau, A. The Future of Global Water Stress: An Integrated Assessment. *Earth's Future* **2014**, *2*, 341–361. [CrossRef]
11. Aulakh, R. China wakes up to its water crisis. *Toronto Star*. 2014. Available online: https://www.thestar.com/news/world/2014/05/12/china_wakes_up_to_its_water_crisis.html (accessed on 18 October 2020).
12. Eikebrokk, B.; Vogt, R.D.; Liltved, H. NOM increase in Northern European source waters: Discussion of possible causes and impacts on coagulation/contact filtration processes. *Water Sci. Technol. Water Supply* **2004**, *4*, 47–54. [CrossRef]
13. Advisian Worley Group. The Cost of Desalination. 2019. Available online: https://www.advisian.com/en/global-perspectives/the-cost-of-desalination (accessed on 18 October 2020).
14. Rotmans, J.; Dowlatabadi, H. Integrated Assessment of Climate Change: Evaluation of Methods and Strategies. In *Human Choices and Climate Change: A State of the Art Report*; Rayner, S., Malone, E.L., Eds.; Tol and Vellinga; Roger Jones Publishing: Harpenden, UK, 1998.
15. Breach, P. *Water Supply Capacity Development in the Context of Global Change*; Electronic Thesis and Dissertation Repository; The University of Western: London, ON, Canada, 2020; Available online: https://ir.lib.uwo.ca/etd/6930 (accessed on 19 October 2020).
16. Breach, P.; Simonovic, S.P. ANEMI 3: Tool for investigating impacts of global change. In *Water Resources Research Report No. 108*; Facility for Intelligent Decision Support, Department of Civil and Environmental Engineering, Eds.; The University of Western Ontario: London, ON, Canada, 2020; 134p, ISBN 978-0-7714-3146-3. Available online: https://www.eng.uwo.ca/research/iclr/fids/publications/products/108.pdf (accessed on 19 October 2020).
17. Davies, E.G.R.; Simonovic, S.P. ANEMI: A new model for integrated assessment of global change. *Interdiscip. Environ. Rev.* **2010**, *11*, 127–161. [CrossRef]

18. Akhtar, M.K.; Wibe, J.; Simonovic, S.P.; MacGee, J. Integrated assessment model of society-biosphere-climate-economy-energy system. *Environ. Model Softw.* **2013**, *49*, 1–21. [CrossRef]
19. Breach, P.; Simonovic, S.P. ANEMI3: An updated tool for global change analysis. *PLoS ONE* **2020**, under review.
20. International Hydrological Programme. *World Freshwater Resources*; UNESCO: Paris, France, 2000; Volume 25, pp. 11–32.
21. Wood, A.; Wollenberg, B. *Power Generation, Operation and Control*, 2nd ed.; Wiley: New York, NY, USA, 1996.
22. Breach, P.A.; Simonovic, S.P. Wastewater Treatment Energy Recovery Potential for Adaptation to Global Change: An Integrated Assessment. *Envir. Manag.* **2018**, *61*, 624–636. [CrossRef] [PubMed]
23. United Nations. *Comprehensive Assessment of the Freshwater Resources of the World*; Stockholm Environment Institute: Stockholm, Sweden, 1997.
24. Davies, E.G.R.; Simonovic, S.P. Global water resources modeling with an integrated model of the social-economic-environmental system. *Adv. Water Resour.* **2011**, *34*, 684–700. [CrossRef]

Publisher's Note: MDPI stays neutral with regard to jurisdictional claims in published maps and institutional affiliations.

© 2020 by the authors. Licensee MDPI, Basel, Switzerland. This article is an open access article distributed under the terms and conditions of the Creative Commons Attribution (CC BY) license (http://creativecommons.org/licenses/by/4.0/).

MDPI

Article

Conflicts in Implementing Environmental Flows for Small-Scale Hydropower Projects and Their Potential Solutions—A Case from Fujian Province, China

Qizhen Ruan [1,2,3], Feifei Wang [1,2,3] and Wenzhi Cao [1,2,3,*]

1 State Key Laboratory of Marine Environmental Science, Xiamen University, Xiamen 361102, China; ruanqizhen@stu.xmu.edu.cn (Q.R.); feifeiw@xmu.edu.cn (F.W.)
2 Key Laboratory of the Ministry of Education for Coastal Wetland Ecosystems, Xiamen University, Xiamen 361102, China
3 College of the Environment & Ecology, Xiamen University, Xiamen 361102, China
* Correspondence: wzcao@xmu.edu.cn

Abstract: Releasing environmental flows is a valuable strategy for mitigating negative impacts of small-scale hydropower projects on river and riparian ecosystems. However, maintaining environmental flows has faced considerable resistance from different stakeholders, and previous studies have failed to appropriately investigate solutions. Here, online questionnaires and interviews were conducted among small-scale hydropower project owners, government administrators, and the public in Fujian Province, China. The results showed that the major hindrance to implementing environmental flows was the potential economic loss resulting from reductions in electricity production, stakeholders' skepticism, technical difficulties, and a lack of the government supervision. Diversion-type projects pose the largest losses of electricity production after the release of environmental flows, and by adopting a 10% of mean annual flow as minimum target, most small-scale hydropower projects obtain low marginal profits without compensation. Here, we proposed an appropriate payment for ecosystem services by introducing an economic compensation program for different types of small-scale hydropower projects scaled by potential losses in electricity generation. Under such a scheme, economic losses from a reduction in electricity production are covered by the government, hydropower project owners, and electricity consumers. Our study offers recommendations for policymakers, officials, and researchers for conflict mitigation when implementing environmental flows.

Keywords: conflicts; environmental flows; small-scale hydropower projects

Citation: Ruan, Q.; Wang, F.; Cao, W. Conflicts in Implementing Environmental Flows for Small-Scale Hydropower Projects and Their Potential Solutions—A Case from Fujian Province, China. *Water* **2021**, *13*, 2461. https://doi.org/10.3390/w13182461

Academic Editors: Athanasios Loukas and Luis Garrote

Received: 8 July 2021
Accepted: 31 August 2021
Published: 7 September 2021

Publisher's Note: MDPI stays neutral with regard to jurisdictional claims in published maps and institutional affiliations.

Copyright: © 2021 by the authors. Licensee MDPI, Basel, Switzerland. This article is an open access article distributed under the terms and conditions of the Creative Commons Attribution (CC BY) license (https://creativecommons.org/licenses/by/4.0/).

1. Introduction

Hydropower is the most common renewable energy source for electricity production. Small-scale hydropower projects (SHPs) play an important role in generating electricity and have been established in 166 countries [1], of which China had ranked first with over 47,498 SHPs by the end of 2017. SHPs in China are defined as having an installed capacity under 50 MW, although there is no internationally agreed definition [2]. The Chinese government encourages the development of renewable energy, such as hydropower and wind power, from which all electricity is purchased by grid companies. There is no unified feed-in tariff for SHPs in China, and each province has the right to set its benchmark price, which is based on SHP development costs and the average purchasing price of the provincial electricity grid company [3].

In the last decade, more attention has been paid to the ecological impacts induced by SHPs, such as hydrological alteration [4–6], river connectivity fragmentation [7], habitat losses [8], and changes in species composition [9,10]. Research has also highlighted the cumulative impacts of SHPs to gain a better understanding of their environmental consequences [11–13]. One important impact is the alteration in natural flow regimes, including river flow depletion [14], which has been proven to be related to the type of hydropower [15–17]. In this study, SHPs were grouped into three categories, namely diversion-type, barrier-type, and mixed-type projects. Both diversion-type and mixed-type projects transfer flow away from natural watercourses through channels or pipelines [18]. Barrier-type projects can be further classified into run-of-river projects and reservoir-type projects depending on their mode of storage. Diversion-type projects are most likely to dry up flows, especially during the dry season [19,20].

Environmental flows (E-flows) refer to discharge volumes that should remain in the river channel [21] to sustain freshwater ecosystem health and human well-being [22]. For the last 50 years, numerous studies have assessed E-flows for ecological health [23–28]. Not until this century have E-flows gradually been incorporated into legislation and regulation practices in many countries [29]. However, in many cases, E-flows are still at the stage of discussion and policy enactment [30], while their implementation faces political, economic, technical, and social challenges [31,32]. We stress that there is a disconnection between booming E-flows science and practice. Currently, there is insufficient work that integrates practice into E-flows literature. Of the existing narratives, there exists a lacuna in mitigating the conflicts in E-flows implementation for SHPs, especially those that incur losses, and the issue of "willingness to pay" [33].

In China, to operate SHPs, one needs an environmental impact assessment and electric power business license. However, E-flows were not involved in environmental impact assessment until the first official requirement of E-flows was stated in 2006 [33]. Additionally, due to neglect of the environmental impacts of SHPs, the regulation only proved effective for large-scale hydropower projects. As the first province in China to enforce E-flows implementation for SHPs, the Fujian provincial government made little progress in implementation until the Jiulong River experienced algal blooming in 2009. This problem was finally solved by opening the sluice gates of all the upper stream hydropower projects. In addition, there are more than 6000 licensed SHPs in Fujian Province [34], including diversion-type (76.7%), barrier-type (11.8%), and mixed-type (11.3%). Crucially, most SHPs in Fujian Province lack the necessary facilities for releasing E-flows because the majority (99.7%) of SHPs had been established before the first Chinese regulation of E-flows was issued.

Discharge and flow velocity are critical factors affecting algal blooming [35,36], which occurs more frequently in rivers with more hydropower projects in Fujian Province. To prevent algal blooming, the Provincial Department of Environmental Protection has required the SHPs of 12 primary rivers to release E-flows and install online monitoring facilities in 2009 [37]. Implementation has involved two different methods, with either "10% of Mean Annual Flow" (10%MAF) [38] or "90-percent exceedance probability of the average flow rate in the driest month based on statistics of monthly mean flows at least 10 years" (Qdm90) as the minimum target [38]. Limited by hydrological data, 10%MAF was used for SHPs in rivers with a drainage area of <500 km^2, which account for 85% of the total SHPs [39], while Qdm90 was adopted by SHPs on the main channels with a drainage area >500 km^2 [40]. However, at the end of 2010, only 28% of the 415 required SHPs had been installed with monitoring facilities [41].

In response to those limitations of the existing literature and urgent demand of releasing E-flows, here, we provide a case study in support of recommendations to facilitate the implementation of E-flows for SHPs. This study is the first known attempt to gather perspectives on E-flows, SHPs, and willingness to pay from three interest groups based on questionnaires and interviews. The objective of the study was to determine the key conflicts in implementing E-flows and to propose potential solutions. By reviewing the literature [42–44], three factors were selected as the main obstacles, namely economics, stakeholders' skepticism, and technologies. Here, we define economic conflicts as economic losses induced by retro-fitting dams and releasing E-flows; stakeholders' skepticism encompasses differences in opinion on whether SHPs are green and the necessity of implementing E-flows; and technical difficulties that include the engineering feasibility of retrofit dams for releasing E-flows. We hypothesized that the threat of economic losses would contribute the most to these potential conflicts, as has been previously suggested in the literature [31,44,45]. We also examine stakeholder perspectives on who should pay for incurred losses and their willingness to pay. Specifically, we aimed to (1) explore the environmental impacts of SHPs in Fujian Province, (2) analyze the difficulties and stakeholder conflicts when implementing E-flows, and (3) examine the current mitigation measures and propose potential solutions.

2. Materials and Methods

2.1. Questionnaires

Online questionnaires were sent to a random sample from members of the public of over 18 years old in Fujian Province. Snowball sampling was adopted to distribute online questionnaires to SHP owners and related government administrators in Fujian Province with the help of the Fujian Province SHP Association by online links because it is recommended for use when samples are rare and difficult to find. There was no preference for selecting respondents in each group. Questionnaires were solely comprised of closed-ended questions with either response at the nominal level or binary response (see Table S1). The sample questions of single choice (A.) and multiple choices (B.) are shown as follows: (A.) Do you think it is necessary to implement environmental flows? (Yes/No); (B.) Who needs to bear the economic loss generated by implementing environmental flows? (The government solely/The owners of SHPs/Electricity consumers by paying more for the electric bill/The government and the owners/The government and electricity consumers/The government, the owners, and electricity consumers).

As it is impossible to know how many times the online questionnaire links had been clicked, we were only able to filter invalid questionnaires by setting up reverse questions; if the obverse and reverse choices were selected at the same time, the questionnaire was considered invalid. The purpose of the study was presented before the questions to ensure each respondent was informed. After a pretest, the question template was re-evaluated; some questions were explained, and some were simplified. The number of questions posed to each target group was different (nine for SHP owners, 11 for government administrators, and six for the general public). All of the questionnaires focused on the environmental impacts of SHPs, attitudes towards SHPs as green enterprises and the E-flows release, perspectives on payment for ecosystem services (PES) as a cost-sharing program, and the willingness to pay for E-flow implementation. The questions to government administrators and owners also covered the conflicts and difficulties of E-flows implementation, average returns and electricity production losses and views on existing compensation policy.

A total of 513 owners, 58 government administrators, and 667 members of public completed the questionnaires, with corresponding validity rates of 93% (478), 93% (55), and 90% (603), respectively. These high validity rates likely reflect the fact that all respondents volunteered to complete the questionnaires, i.e., people with a low willingness to respond would ignore the original links. The chi-square test was adopted to examine the differences in the choices of respondent groups, where $p < 0.05$ indicated a significant difference.

Yet, respondent accessibility has the potential to affect the response rate and prejudice the results, particularly in a survey targeted at a large area. Other survey limitations include gathering responses from those who did not actively participate.

2.2. Interviews

Semi-structured interviews were conducted after the questionnaires were collected to obtain a more comprehensive understanding of the perspectives of different interest groups. These semi-structured interviews were flexible and allowed interviewers to alter the pace and order of questions depending on interviewees to acquire their best responses. The interviewees were communicated via social media, informed about the purpose of the study, and asked if they would be willing to participate. Three SHP owners, a county government administrator, a provincial government administrator, as well as a hydro-ecology engineer agreed to join. The three SHP owners and the administrator were involved in pilot work for this project in 2014. One owner had previously retrofitted his facility with sluices and installed an ecological generator, and the other two owners had their projects decommissioned. The administrator had long-term experience in SHP management. The hydro-ecology engineer was selected as a representative member of the public. Interviews were recorded when permitted. The semi-interviews lasted 20–40 min and covered details of the SHPs of the interviewed owners as well as attitudes and perspectives on the challenges of E-flows implementation. The questions for SHP owners, the administrator and the engineer consisted of 6 close-ended questions and 3 open-ended questions; 3 close-ended questions and 3 open-ended questions; and 2 close-ended questions and 4 open-ended questions, respectively. Those questions covered attitudes and perspectives on the challenges of E-flows implementation and the issue of "who needs to pay for the losses" (see Tables S2–S4).

2.3. Secondary Data Collection

Secondary sources were used to determine environmental impacts as well as the losses in electricity production caused by E-flows implementation and the average returns of SHPs. Sources included the SHP Annual Statistical Report (2016) in Fujian Province and the 2017 Survey report on the status of rural hydropower projects in Fujian Province. Information and data were also obtained from the Fujian Provincial Department of Water Resources.

3. Results

3.1. Environmental Impacts

The results from the questionnaires showed that 20% of government administrators had once received letters of complaint related to dry watercourses caused by SHPs from local residents. This ecological impact was evidenced by government reports, with more than 93% (5815) of the projects resulting in dry reaches accounting for a total length of 7508.5 km, and around 7% (430) of the projects cut flows of up to at least 3 km of dry reaches (Figure 1).

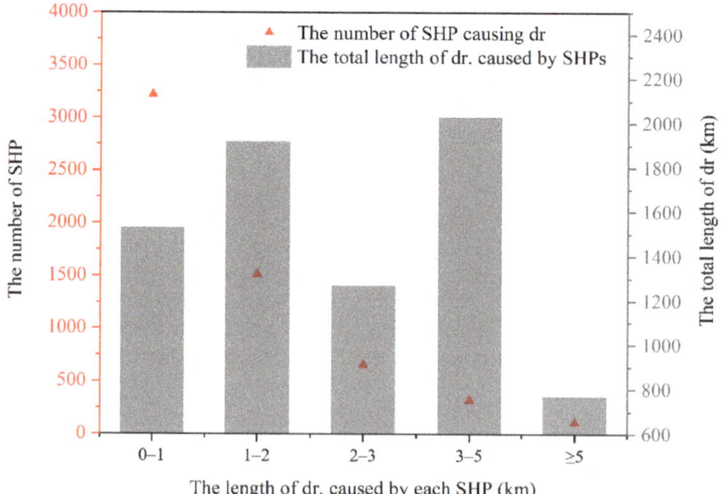

Figure 1. Lengths of dry reaches (dr.) generated by small-scale hydropower projects in Fujian Province. Source: [46].

In addition to dry river reaches, the excessive construction of SHPs in Fujian Province had turned some river reaches into reservoirs. Data collected from the Fujian Provincial Department of Water Resources show an average SHP spacing of 13 km on 65 rivers with a drainage area > 500 km^2, which has resulted in large decreases in discharge and flow velocity, which is conducive to algae growth. For example, on the Jiulong River, 10 SHPs operate on the trunk reach with an average spacing of <7 km and the smallest spacing of just 5.4 km.

3.2. Conflicts

The initial attempt to implement E-flows regulation in Fujian Province encountered much resistance, with economic factors identified as the main obstacle, followed by stakeholders' skepticism and technical difficulties (Figure 2).

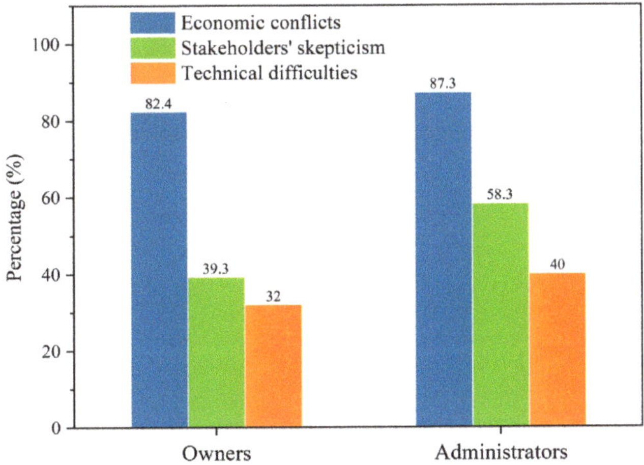

Figure 2. Opinions of small-scale hydropower project owners and government administrators on the difficulties of implementing E-flows (select one or more answer choices).

3.2.1. Stakeholders' Skepticism

All respondent groups tended to be positive about "whether SHPs are green energy", with support from 96.4% of the owners, 90.9% of the administrators, and 81.8% of the public. However, of the corresponding groups, 51.7%, 69.1%, and 81.9% considered SHPs to have negative impacts on the environment, respectively. Similar proportions (54.6%, 63.3%, and 91%, respectively) supported the implementation of E-flows.

3.2.2. Economic Conflicts

Because the administrators required the SHPs to release E-flows without any compensation, SHP owners were not willing to follow the requirement. Implementing E-flows involves reducing the discharge volumes available to produce electricity, which inevitably results in economic losses for SHP owners.

The SHP owner and SHP administrator groups were asked about the magnitude of losses experienced with the 10%MAF strategy, the results of which are shown in Figure 3. Nearly two-thirds of the owners of diversion-type SHP and nearly half of the owners of barrier- and mixed-type SHP suggested that their losses would exceed 10%. The diversion-type SHP owners estimated losses to be more than 15%, and these estimates were much higher than those of the barrier- and mixed-type SHP owners.

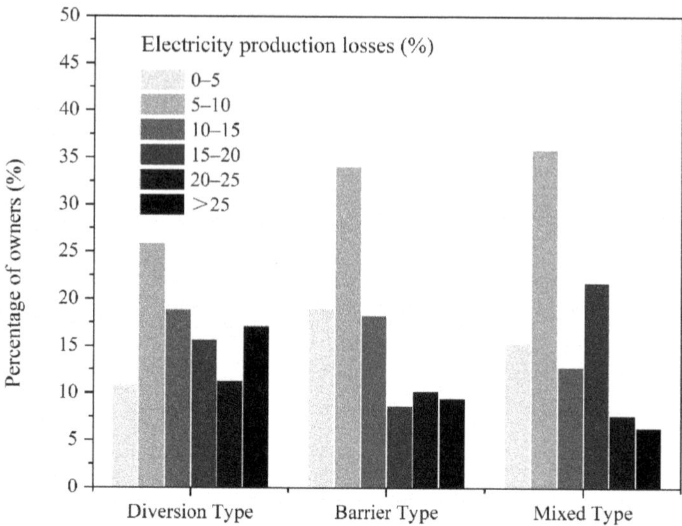

Figure 3. Influences of releasing E-flows on electricity production losses estimated by the owners of different types of small-scale hydropower projects.

These estimates are corroborated by our calculations of electricity production losses assuming the 10%MAF method based on available information, which implies that losses vary between SHPs types (Table 1). For example, impacts on diversion-type projects ranged between 9.7% and 23.6%. Overall, the calculated losses tended to decrease with SHP capacity; for barrier-type projects, production losses ranged from 3.6% to 8.6% and decreased in line with the single installed capacity of the SHPs. In general, reservoir-type projects only have generators with large capacities, while run-of-river projects usually have more generators with different installed capacities. Therefore, reservoir-type projects typically suffer comparatively higher production losses. The losses of mixed-type projects vary from 9.7% to 11.7% of their expected electricity production. Both official data (Table 1) and the questionnaire responses show that different types of the SHPs are subject to varying production losses as a result of E-flows regulation, with diversion-type (accounting for 76.7% of the projects in Fujian Province) being most affected.

An analysis of the questionnaire responses from SHP owners and administrators relating to estimates of electricity production losses and average returns is presented in Figure 4. The SHP owners estimated slightly higher losses than the administrators, and more than one-third and one-fifth of the owners and administrators believed that E-flows accounted for 15% of losses, respectively. One-third of the administrators believed the losses were low (0–5%), while only one-seventh of the owners believed this was the case (Figure 4a). Overall, the administrators were more optimistic than the owners, with approximately one-third believing that SHP received > 10% profit compared to one-fifth of the owners (Figure 4b). Despite differences in opinions on average returns of SHPs, there was no marked difference in estimates of electricity production losses ($p > 0.05$), with both groups suggesting relatively losses overall (Figure 4a).

Table 1. Influence of releasing E-flows on electricity production losses of small-scale hydropower projects. Source: [47].

The Name of SHPs	SHP Type	Catchment Area	Installed Capacity	Reservoir Storage	10%MAF	Electricity Production Losses
		km^2	MW	10^4 m^3	m^3/s	%
United Huiji	Diversion type	63	0.50	247.0	0.182	12.6
Lutouxia	Diversion type	285	2.50	49.1	0.820	16.4
Dongxiwei	Diversion type	5	0.25	3.0	0.0013	23.6
Yongxi	Diversion type	224	40.00	6900.0	0.795	9.7
Longmeishan	Diversion type	22	1.00	14.5	0.060	16.8
Sixth Cascade Project of Qingyin River	Diversion type	329	7.500	4400.0	0.930	10.9
Shanzai	Barrier type (Reservoir type)	1646	33.00	17,600.0	5.700	8.6
Shangjishan	Barrier type (Run-of-river)	1138	2.80	61.0	3.570	3.6
Dongxi	Mixed type	42	3.20	189.9	0.215	11.7
Yangmeizhou	Mixed type	128	11.30	201.0	0.253	11.5
Fuquanxi I	Mixed type	116	8.50	1758.0	0.380	9.7
Fuquanxi II	Mixed type	158	36.10	337.0	0.730	10.16

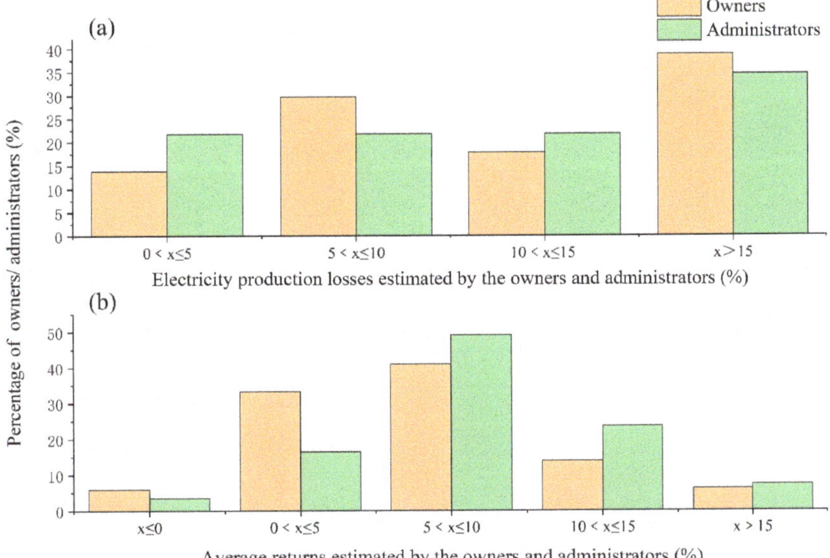

Figure 4. Estimates of (**a**) electricity production losses overall and (**b**) average returns of small-scale hydropower projects by owners and administrators.

It is known that profits of SHPs will be lowered to several levels after releasing E-flows, particularly for diversion-type SHPs. Because the SHPs have E-flows releasing infrastructure, most SHP owners have to pay for retrofit costs. Without subsidy, the costs of releasing E-flows could only be covered by the owners, and this would worsen conflicts, resulting in their unwillingness to follow the regulations.

3.2.3. Other Difficulties

In addition to economic issues and perspective divergences, E-flows policies are impeded by technical difficulties and management weaknesses. Although technical issues ranked last amongst the potential challenges (Figure 2), the installation of online monitoring facilities remains technically difficult. For example, wireless signals are poor in remote areas, making it impossible to perform online network monitoring. In the case of governmental management, 66% of the public did not believe that owners would release E-flows without strict government supervision even if they were given proper compensation.

3.3. *Approaches to Mitigating Conflicts*

3.3.1. Pilot Approaches

Due to the failure of the initial E-flows policy, the Fujian Provincial Department of Water Resources selected Changting and Yongchun counties as pilot sites to explore PES approaches to mitigating conflicts. In this program, SHP owners obtained extra on-grid tariff subsidies or one-off compensation payments based on the following approaches according to interview responses for illustration.

i. Retrofit works

Existing facilities were retrofitted to release E-flows, such as adding flow release holes and modifying sluices. The installation of ecological generators was also encouraged, which can utilize E-flows to generate electricity to reduce economic losses.

Xiyuan projects included a reservoir- and a diversion-type project, the latter being 1.5 km downstream of the former along a 2-km diversion channel and a 1.5-km dry reach. The Xiyuan reservoir-type project was later retrofitted by adding sluices in the diversion channel to release E-flows, and an ecological generator with a capacity of 0.125 MW was installed in the power plant of the reservoir project. The estimated losses caused by releasing E-flows (340,000 kWh) equate to approximately 0.095 million CNY, equivalent to an extra 0.07 CNY/kWh (on-grid tariff) to cover the losses. The additional cost for the ecological generator, sluice retrofit, and monitoring facilities was 0.46 million CNY. The government offers an extra on-grid tariff of 0.05 CNY/kWh as compensation as well as a subsidy of about 50% of the cost of the ecological generator.

ii. Restricted seasonal operation

These diversion-type projects, which cannot meet the E-flows needs, were prohibited from operating during the dry season (December–February). For example, the Qingyuan diversion-type SHP, which resulted in a 6.8-km-dry reach, was prohibited from running during the season. The estimated resulting production losses were 0.31 million kWh, equivalent to approximately 0.93 million CNY, with an additional on-grid tariff of 0.072 CNY/kWh needed to cover the loss. The cost of the sluice retrofit and monitoring facilities also exceeded 0.05 million CNY. Based on the PES scheme, the SHP owner received an extra on-grid tariff of 0.05 CNY/kWh as compensation.

iii. Decommissioning

SHPs that are too difficult to retrofit were decommissioned under the condition of guaranteed irrigation and public safety.

The downstream section of the Hongqi SHP area is a popular natural spot. However, due to the improper operation of the project, flows in the trunk stream were delivered to diversion channel, leading to a 2.4-km dry reach and significant damage to the landscape character. After prolonged negotiation, 2.52 million CNY (approximately 60% of the appraisal price of the project considering the installed capacity, electricity production, on-grid tariff, construction time, etc.) was paid as compensation for dismantling the project. The Hongqi project dam was eventually removed, although its power plant was retained as a hydropower museum.

Such practices have been successful in the study region by addressing the occurrence of dry reaches caused by SHPs. It is noted that instead of being dismantled, some powerplants have been converted into museums, cafés, or libraries, thereby providing beneficial public spaces for the neighboring communities. "The government's regulations" and "the obligation and responsibilities for the environment" were all mentioned by the interviewed SHP owners when asked why they finally agreed to implement E-flows or decommission their projects. Those owners who supported the implementation of E-flows emphasized that to avoid opposition, the government ought to fully consider stakeholders' interest. In addition to compensation and subsidies, government administrators also attributed the success of these schemes to "constant communication and negotiation between the government administrators and the owners" as well as "the long-term publicity towards the importance of ecosystems".

3.3.2. Current Economic Incentives

In light of the success of the pilot PES scheme, the Fujian Provincial government required all SHPs to be installed with monitoring facilities by the end of 2020 [48]. To facilitate this, SHP owners receive various levels of compensation depending on the nature of the work, i.e., retrofit work, seasonally restricted operation, or decommissioning. The projects requiring retrofitting work are awarded an extra on-grid tariff of 0.02 CNY/kWh; those adopting seasonally restricted operation are subsidized by an extra 0.03 CNY/kWh; and for decommissioned projects, owners can receive 50% of the market price as compensation. By the end of 2019, 1966 projects had already implemented E-flows, and 584 projects had been decommissioned [49]. Because the energy supply in Fujian Province is sufficient, the losses in electricity production resulting from these schemes do not currently have any negative consequences for industrial production or the standard of living.

To date, retrofit works have been widely applied, although the owners of diversion-type projects have suffered relatively higher losses than those of barrier-type projects. Therefore, the fairness of the different PES schemes may become an issue. Based on the questionnaires, SHP owners adopting seasonally restricted operations did not consider the PES subsidy sufficient, with only 39.5% of the owners and 21.8% of the administrators supporting this scheme. Indeed, only approximately 10% of the SHP owners and administrators were satisfied with current economic incentives, while approximately half of these two groups expected incentives to be scaled based on relative economic losses.

4. Discussion

4.1. Improving PES Programs

Individuals adversely affected by environmental policies need to be sufficiently compensated [50]. In the case of E-flows, relatively low levels of compensation will influence the sustainability of policies, as although owners may reluctantly release E-flows in the short run under pressure from the government, less effort will be given to maintenance and management of E-flows over the longer term.

Furthermore, the calculation of E-flow impacts varies depending on which methods of assessment is adopted. Furthermore, losses in electricity production vary between SHPs —even under the same hydrological circumstances—depending on which approaches are taken [51]. However, current compensation strategies do not reflect the actual losses incurred by owners. Thus, it is more reasonable to apply differentiated compensation based on SHP electricity generation losses rather than E-flows. Therefore, a price system based on differential compensation according to the actual electricity production losses incurred to maintain E-flows is recommended.

Apart from the amount of compensation [52], the source of compensation is a prickly issue to tackle. A long-term, funding-supported system should be established as soon as possible [29]. The benefits of restoring river ecosystems are well known and all beneficiaries must bear some responsibility. As the direct parties involved, SHP owners should take the initiative to undertake environmental improvements, whereas consumers, as indirect parties, need to take responsibility for triggering the demand for environmental services. Therefore, both parties should bear some of the losses caused by implementing E-flows. The government can offer subsidies for retrofitting dams, installing ecological generators, and monitoring facilities. Considering that government finances may not be able to afford ongoing compensation, we propose a cost-sharing PES program paid by all interest groups. Similar to thermal power, for which on-grid tariff includes the costs of denitration, there is an opportunity to recover the partial costs of releasing E-flows from some electricity consumers. The raised on-grid tariff of hydropower was still the lowest among all types of energy in Fujian Province, at 0.33 CNY/kWh compared to 0.39 CNY/kWh for thermal power [53], 0.4 CNY/kWh for nuclear power [54], and 0.48 CNY/kWh for wind power [55]. Based on our questionnaires, the option of sharing the additional costs of implementing E-flows between government, owners, and electricity consumers gained the highest level of support among each group (Figure 5), which suggests the potential for establishing such a PES program.

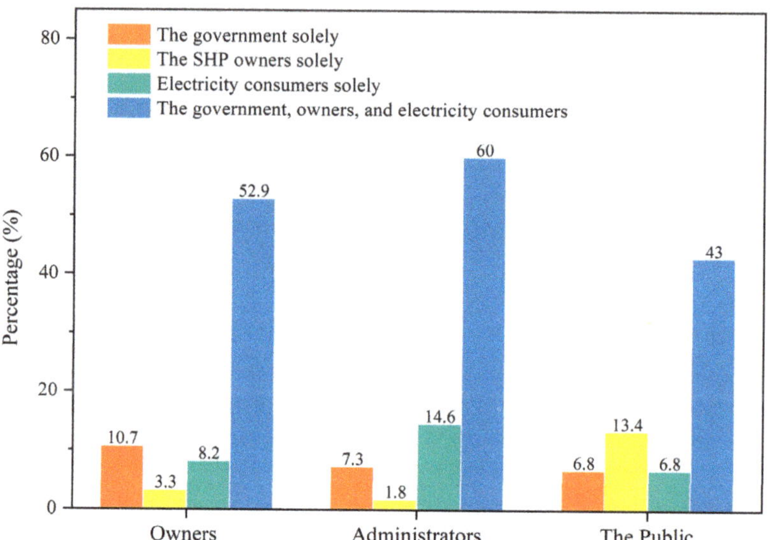

Figure 5. Comparison of the views of short-term hydropower project owners, administrators, and the public on a cost-sharing PES program for E-flows (select one or more answer choices).

Based on the electricity generation by different types of SHPs, if all SHPs adopt the 10%MAF strategy when implementing E-flows, the average electricity production losses are estimated to be approximately 9.38% of the total electricity production of SHPs in Fujian Province. Considering the average annual electricity generation (approximately 23.3 billion kWh [56]) and on-grid tariff of SHP in Fujian Province, the losses of releasing E-flows are calculated as 2.18 billion kWh, which amounts to approximately 721.22 million CNY. According to the total electricity consumption (211.27 billion kWh [57]) and the average on-grid tariff (0.6 CNY/kWh) in Fujian Province, the total electricity consumption costs 126.762 billion CNY. The impact ratio of electricity bills is derived from losses of releasing E-flows divided by total electricity consumption costs, which is calculated as 0.57%.

If all the losses caused by implementing E-flows are transferred to the electricity rate, the impact on people's original electricity bills is approximately 0.56%. On the basis of the cost-sharing principle, electricity consumers and the SHP owners would bear this impact together; if this system was adopted, electricity consumers pay < 0.56% more than their existing electricity bills. The results from the questionnaires show that although there is significantly less support from administrators than the public, nearly three-quarters of both groups were willing to pay 1% more than their usual electricity bills to support E-flow implementation.

4.2. *Improvement of Communication and Management*

The cooperation of stakeholders is essential for successful E-flows implementation. In general, the greater the acceptance of the need for E-flows, the more likely a successful partnership can be formed. Our results indicate that the perception of the necessity of E-flows differs between and within interest groups. Thus, all groups need to increase their eco-awareness of the need to achieve environmental protection. Ensuring appropriate that communication during the decision-making process will further ensure the success of implementing scheme [58]. This is crucial to enable all stakeholders to raise and resolve potential disagreements [59]. Our study showed that communication is an essential component of collaboration; active dialog between interest groups helps to reach a compromise, allowing potential conflicts to the recognized and addressed during the implementation process.

Moreover, understanding of SHPs is often one-sided, largely depending on where their benefits lie. SHPs are generally welcomed, as they are the cheapest and most accessible means of obtaining electricity [60,61]; however, with improving living standards, Chinese residents have begun to pay more attention to environmental quality. Indeed, E-flows schemes have little negative impact on the public's economic interest but bring environmental and recreational benefits, which may account for their relatively low level of recognition of "SHPs belong to green energy" and high level of support for E-flows implementation.

Monitoring the long-term impacts of current measures is also helpful for informing subsequent management [44]. As there remain unknown relationships between flows and biotic responses [62], monitoring is needed to address this uncertainty [63], and local electricity users can be successfully involved in this monitoring work [64]. Additionally, publishing the outcomes of current monitoring measures should help bolster public support [65], which would likely enhance public desire for further E-flows implementation. Specific E-flow assessments could be conducted on SHPs located in ecologically sensitive regions in light of the capacity and available resources of regional and local governments. Undoubtedly, gradually augmenting the scale of E-flows implementation seems inevitable, which must be matched by suitable compensation schemes.

As people's environmental requirements have changed, government understanding and regulation of water resources need to change too [16]. Future water resources planning should strive for both comprehensive and coordinated development of the environment and society. Taking environmental factors into account at the planning stage will help identify potential stakeholder conflicts that will otherwise need to be tackled at a later date.

5. Conclusions

E-flows have been recognized as a crucial water management tool when aiming to meet both environmental and societal needs. This study represents, to the best of our knowledge, the first attempts at exploring solutions to mitigate the conflicts in E-flows implementation for SHPs based on questionnaires and interviews of three interest groups. We used Fujian Province as a case study to demonstrate the challenges facing E-flow implementation, focusing on (1) skepticism about "whether SHPs are green" and "the necessity of releasing E-flows" among SHP owners, government administrators, and the general public; (2) economic conflicts caused by electricity production losses especially in the case of diversion-type projects; (3) inadequate governance; and (4) PES. Importantly, our questionnaires and interviews reveal that there is potential for establishing a long-term cost-sharing PES program, paid by the government, SHP owners, and electricity consumers and emphasize that successful E-flows implementation will benefit from sustained and effective communication between all interest groups.

As E-flows enter the implementation phase, it should be recognized that economic challenges remain the strongest driver and key obstacle to implementing environmental policies [66]. Furthermore, it is worth recognizing that while E-flows implementation is a valuable tool, this is not the only measure available for river rehabilitation concerned with SHPs. For example, fish pass facilities need to be established to improve longitudinal continuity. While beyond the scope of this study, further work is also needed to consider the ecological responses to E-flows schemes so that they can be enhanced and optimized in the future. This requires the cooperation of scientists and water managers [67]. Finally, we emphasize that a combination of social, economic, and environmental disciplines is needed to enhance existing understanding and overcome the potential challenges of implementing and managing E-flows schemes.

Supplementary Materials: The following are available online at https://www.mdpi.com/2073-4441/13/18/2461/s1, Table S1: Questionnaires sent to SHP owners, SHP government administrators, and the public; Table S2: Interview protocol questions for SHP owners; Table S3: Interview protocol questions for SHP government administrator; Table S4: Interview protocol questions for hydro-ecology engineer.

Author Contributions: Conceptualization, Q.R.; methodology, Q.R.; validation, F.W.; investigation, Q.R.; resources, Q.R.; writing—original draft preparation, Q.R.; writing—review and editing, W.C.; supervision, W.C.; project administration, W.C. and F.W. All authors have read and agreed to the published version of the manuscript.

Funding: This research was funded by the National Natural Science Foundation of China (41771500) and the Environmental Protection Technology Program of Fujian Province (2021R023).

Institutional Review Board Statement: Not applicable.

Informed Consent Statement: Informed consent was obtained from all subjects involved in the study.

Data Availability Statement: Data are contained within the article and Supplementary Material.

Acknowledgments: Our sincere thanks go to survey respondents for patient answers, to Nina Morris from the University of Edinburgh, and Baoli Liu from Fuzhou University for their valuable advice.

Conflicts of Interest: The authors declare no conflict of interest.

References

1. United Nations Industrial Development, the International Center on Small Hydro Power. World Small Hydropower Development Report 2019. 2019. Available online: www.smallhydroworld.org (accessed on 2 August 2021).
2. Couto, T.B.A.; Olden, J.D. Global proliferation of small hydropower plants—Science and policy. *Front. Ecol. Environ.* **2018**, *16*, 91–100. [CrossRef]
3. Ministry of Water Resources, Rural Hydropower, Electrification Bureau. *The 2017 Rural Hydropower Statistics Yearbook*; China Statistics Press: Beijing, China, 2018.
4. Ibisate, A.; Díaz, E.; Ollero, A.; Acín, V.; Granado, D. Channel response to multiple damming in a meandering river, middle and lower Aragón River (Spain). *Hydrobiologia* **2013**, *712*, 5–23. [CrossRef]
5. Lu, W.; Lei, H.; Yang, D.; Tang, L.; Miao, Q. Quantifying the impacts of small dam construction on hydrological alterations in the Jiulong River basin of Southeast China. *J. Hydrol.* **2018**, *567*, 382–392. [CrossRef]
6. Xiao, X.; Chen, X.; Zhang, L.; Lai, R.; Liu, J. Impacts of small cascaded hydropower plants on river discharge in a basin in Southern China. *Hydrol. Process.* **2019**, *33*, 1420–1433. [CrossRef]
7. Wang, D.; Wang, X.; Huang, Y.; Zhang, X.; Zhang, W.; Xin, Y.; Qu, M.; Cao, Z. Impact analysis of small hydropower construction on river connectivity on the upper reaches of the great rivers in the Tibetan Plateau. *Glob. Ecol. Conserv.* **2021**, *26*, e01496. [CrossRef]
8. Robert, A. *River Processes: An Introduction to Fluvial Dynamics*; Robert, A., Ed.; Arnold: London, UK, 2003; ISBN 0340763396.
9. Mueller, M.; Pander, J.; Geist, J. The effects of weirs on structural stream habitat and biological communities. *J. Appl. Ecol.* **2011**, *48*, 1450–1461. [CrossRef]
10. Benejam, L.; Saura-Mas, S.; Bardina, M.; Solà, C.; Munné, A.; García-Berthou, E. Ecological impacts of small hydropower plants on headwater stream fish: From individual to community effects. *Ecol. Freshw. Fish* **2016**, *25*, 295–306. [CrossRef]
11. Deitch, M.J.; Merenlender, A.M.; Feirer, S. Cumulative effects of small reservoirs on streamflow in Northern Coastal California catchments. *Water Resour. Manag.* **2013**, *27*, 5101–5118. [CrossRef]
12. Kibler, K.M.; Tullos, D.D. Cumulative biophysical impact of small and large hydropower development in Nu River, China. *Water Resour. Res.* **2013**, *49*, 3104–3118. [CrossRef]
13. Kuriqi, A.; Pinheiro, A.N.; Sordo-Ward, A.; Bejarano, M.D.; Garrote, L. Ecological impacts of run-of-river hydropower plants—Current status and future prospects on the brink of energy transition. *Renew. Sustain. Energy Rev.* **2021**, *142*, 110833. [CrossRef]
14. Bakken, T.H.; Sundt, H.; Ruud, A.; Harby, A. Development of Small Versus Large Hydropower in Norway–Comparison of Environmental Impacts. *Energy Procedia* **2012**, *20*, 185–199. [CrossRef]
15. Egré, D.; Milewski, J.C. The diversity of hydropower projects. *Energy Policy* **2002**, *30*, 1225–1230. [CrossRef]

16. Zhang, J.; Xu, L.; Yu, B.; Li, X. Environmentally feasible potential for hydropower development regarding environmental constraints. *Energy Policy* **2014**, *73*, 552–562. [CrossRef]
17. Kelly-Richards, S.; Silber-Coats, N.; Crootof, A.; Tecklin, D.; Bauer, C. Governing the transition to renewable energy: A review of impacts and policy issues in the small hydropower boom. *Energy Policy* **2017**, *101*, 251–264. [CrossRef]
18. Gower, T.; Rosenberger, A.; Peatt, A.; Hill, A. *Tamed Rivers: A Guide to River Diversion Hydropower in British Columbia*; Watershed Watch Salmon Society: Vancouver, BC, Canada, 2012.
19. Hennig, T.; Wang, W.; Feng, Y.; Ou, X.; He, D. Review of Yunnan's hydropower development. Comparing small and large hydropower projects regarding their environmental implications and socio-economic consequences. *Renew. Sustain. Energy Rev.* **2013**, *27*, 585–595. [CrossRef]
20. Kumar, D.; Katoch, S.S. Sustainability suspense of small hydropower projects: A study from western Himalayan region of India. *Renew. Energy* **2015**, *76*, 220–233. [CrossRef]
21. Operacz, A.; Wałęga, A.; Cupak, A.; Tomaszewska, B. The comparison of environmental flow assessment-the barrier for investment in Poland or river protection? *J. Clean. Prod.* **2018**, *193*, 575–592. [CrossRef]
22. The Brisbane Declaration. Available online: http://riverfoundation.org.au/wp-content/uploads/2017/02/THE-BRISBANE-DECLARATION.pdf (accessed on 4 September 2021).
23. Ward, J.V. *The Ecology of Regulated Streams*; Ward, J.V., Stanford, J.A., Eds.; Plenum Press: New York, NY, USA; London, UK, 1979.
24. Caldwell, P.V.; Kennen, J.G.; Sun, G.; Kiang, J.E.; Butcher, J.B.; Eddy, M.C.; Hay, L.E.; LaFontaine, J.H.; Hain, E.F.; Nelson, S.A.C.; et al. A comparison of hydrologic models for ecological flows and water availability. *Ecohydrology* **2015**, *8*, 1525–1546. [CrossRef]
25. Prakasam, C.; Saravanan, R.; Kanwar, V.S. Evaluation of environmental flow requirement using wetted perimeter method and GIS application for impact assessment. *Ecol. Indic.* **2021**, *121*, 107019. [CrossRef]
26. Tharme, R.E. A global perspective on environmental flow assessment: Emerging trends in the development and application of environmental flow methodologies for rivers. *River Res. Appl.* **2003**, *19*, 397–441. [CrossRef]
27. Linares, M.S.; Assis, W.; Castro Solar, R.R.; Leitão, R.P.; Hughes, R.M.; Callisto, M. Small hydropower dam alters the taxonomic composition of benthic macroinvertebrate assemblages in a neotropical river. *River Res. Appl.* **2019**, *35*, 725–735. [CrossRef]
28. Arnett, J.L. *Methodologies for the Determination of Stream Resource Flow Requirements: An Assessment*; Stalnaker, C.B., Ed.; Utah State University: Utah, UT, USA, 1976.
29. Le Quesne, T.; Kendy, E.; Weston, D. *The Implementation Challenge: Taking Stock of Government Policies to Protect and Restore Environmental Flows*; WWF: Gland, Switzerland, 2010.
30. Hirji, R.; Davis, R. *Environmental Flows in Water Resources Policies, Plans, and Projects: Case Studies*; The World Bank: Washington, DC, USA, 2009; ISBN 978-0-8213-7940-0.
31. Pérez-Díaz, J.I.; Wilhelmi, J.R. Assessment of the economic impact of environmental constraints on short-term hydropower plant operation. *Energy Policy* **2010**, *38*, 7960–7970. [CrossRef]
32. Harwood, A.; Johnson, S.; Richter, B.; Locke, A.; Yu, X.; Tickner, D. *Listen to the River: Lessons from a Global Review of Environmental Flow Success Stories*; WFF: UK, 2017. Available online: http://indiaenvironmentportal.org.in/files/file/listen_to_the_river.pdf (accessed on 7 September 2021).
33. Kataria, M. Willingness to pay for environmental improvements in hydropower regulated rivers. *Energy Econ.* **2009**, *31*, 69–76. [CrossRef]
34. Fujian Provincial Department of Water Resources. *The 2017 Annual Statistical Report of Small-Scale Hydropower Projects in Fujian Province*; Fujian Provincial Department of Water Resources: Fuzhou, China, 2018.
35. Mitrovic, S.M.; Hardwick, L.; Dorani, F. Use of flow management to mitigate cyanobacterial blooms in the Lower Darling River, Australia. *J. Plankton Res.* **2011**, *33*, 229–241. [CrossRef]
36. Liu, X.; Li, Y.-L.; Liu, B.-G.; Qian, K.-M.; Chen, Y.-W.; Gao, J.-F. Cyanobacteria in the complex river-connected Poyang Lake: Horizontal distribution and transport. *Hydrobiologia* **2016**, *768*, 95–110. [CrossRef]
37. Fujian Provincial Economic and Trade Commission, Fujian Provincial Department of Environmental Protection, Fujian Provincial Department of Water Resources, Fujian Energy Regulatory Office. *Letter on Further Improving the Installation of Online Monitoring Devices for Implementing Environmental Flows of Hydropower Projects*; Fujian Provincial Economic and Trade Commission, Fujian Provincial Department of Environmental Protection, Fujian Provincial Department of Water Resources, Fujian Energy Regulatory Office: Fuzhou, China, 2009.
38. Ministry of Water Resources. *Technical Specification for Analysis of Supply and Demand Balance of Water Resources*; China Water & Power Press: Beijing, China, 2008; SL 429-2008.
39. Fujian Provincial Department of Water Resources. *Hydro Resources Planning for the Small and Medium Watershed in Fujian Province*; Fujian Provincial Department of Water Resources: Fuzhou, China, 2015.
40. Fujian Provincial Department of Water Resources. *Integrated Planning for the Large Watershed in Fujian Province*; Fujian Provincial Department of Water Resources: Fuzhou, China, 2009.
41. Fushui, R. The practices and perspectives of the ecological restoration and rehabilitation Affected by Small-scale Hydropower Stations. *Water Resour. Dev. Res.* **2017**, *17*, 35–38.
42. Moore, M. Perceptions and Interpretations of Environmental Flows and Implications for Future Water Resource Management: A Survey Study. Master's Thesis, Linköping University, Norrköping, Sweden, 2004.

43. Dyson, M.; Bergkamp, G.; Scanlon, J. *Flow: The Essentials of Environmental Flows*; IUCN: Gland, Switzerland; Cambridge, UK, 2003; ISBN 2-8317-0725-0.
44. Harwood, A.J.; Tickner, D.; Richter, B.D.; Locke, A.; Johnson, S.; Yu, X. Critical factors for water policy to enable effective environmental flow implementation. *Front. Environ. Sci.* **2018**, *6*, 37. [CrossRef]
45. Guisández, I.; Pérez-Díaz, J.I.; Wilhelmi, J.R. Approximate formulae for the assessment of the long-term economic impact of environmental constraints on hydropeaking. *Energy* **2016**, *112*, 629–641. [CrossRef]
46. Fujian Provincial Department of Water Resources. *SHP Annual Statistical Report 2016 in Fujian Province*; Fujian Provincial Department of Water Resources: Fuzhou, China, 2017.
47. Fujian Provincial Department of Water Resources. *The 2017 Survey Report on the Status of Rural Hydropower Projects in Fujian Province*; Fujian Provincial Department of Water Resources: Fuzhou, China, 2017.
48. Fujian Provincial Pricing Administration, Fujian Provincial Economic and Trade Commission & Fujian Provincial Department of Environmental Protection, Fujian Provincial Department of Water Resources. *System of ecological on-grid tariff for hydropower stations in Fujian Province (Trial Implementation)*; Fujian Provincial Pricing Administration, Fujian Provincial Economic and Trade Commission & Fujian Provincial Department of Environmental Protection, Fujian Provincial Department of Water Resources: Fuzhou, China, 2017.
49. Fujian Provincial Department of Water Resources. *The Report on Fujian Provincial Department of Water Resources on the Environmental Flows of Rural Hydropower Projects*; Fujian Provincial Department of Water Resources: Fuzhou, China, 2020.
50. Aylward, B.; Emerton, L. Building capacity for design and implementation. In *Flow: The Essentials of Environmental Flows*; Dyson, M., Bergkamp, G., Scanlon, J., Eds.; IUCN: Gland, Switzerland; Cambridge, UK, 2003; pp. 31–43, ISBN 2-8317-0725-0.
51. Bejarano, M.D.; Sordo-Ward, A.; Gabriel-Martin, I.; Garrote, L. Tradeoff between economic and environmental costs and benefits of hydropower production at run-of-river-diversion schemes under different environmental flows scenarios. *J. Hydrol.* **2019**, *572*, 790–804. [CrossRef]
52. Sisto, N.P. Environmental flows for rivers and econocmic compensation for irrigators. *J. Environ. Manag.* **2009**, *90*, 1236–1240. [CrossRef] [PubMed]
53. Fujian Provincial Development and Reform Commission. *Notice of the Implementation Plan for Deepening the Reform of Coal-Fired Power Generation in Fujian Province*; Fujian Provincial Development and Reform Commission: Fuzhou, China, 2020.
54. Fujian Provincial Pricing Administration, Fujian Provincial Economic and Trade Commission, Fujian Provincial Department of Environmental Protection. *Notice of Adjusting on-Grid Tariffs for Nuclear Power*; Fujian Provincial Pricing Administration, Fujian Provincial Economic and Trade Commission, Fujian Provincial Department of Environmental Protection: Fuzhou, China, 2017.
55. National Development and Reform Commission. *Notice on Improving Wind Power Tariff Policy*; National Development and Reform Commission: Beijing, China, 2019.
56. China Business Industry Research Institute. 2021–2026 Fujian Informatization Market Survey and Development Research Report. 2020. Available online: https://www.askci.com/reports/20210202/1432081583895421.shtml (accessed on 7 September 2021).
57. State Grid Fujian Electric Power Company. 2017 Annual Report on Power Supply Information. 2018. Available online: http://www.fj.sgcc.com.cn/html/main/col2841/2019-11/22/20191122145256479936120_1.html (accessed on 7 September 2021).
58. King, J.; Acreman, M. Covering the cost. In *Flow: The Essentials of Environmental Flows*; Dyson, M., Bergkamp, G., Scanlon, J., Eds.; IUCN: Gland, Switzerland; Cambridge, UK, 2003; pp. 31–43, ISBN 2-8317-0725-0.
59. Pahl-Wostl, C.; Arthington, A.; Bogardi, J.; Bunn, S.E.; Hoff, H.; Lebel, L.; Nikitina, E.; Palmer, M.; Poff, L.N.; Richards, K.; et al. Environmental flows and water governance: Managing sustainable water uses. *Curr. Opin. Environ. Sustain.* **2013**, *5*, 341–351. [CrossRef]
60. Kelly, S. Megawatts mask impacts: Small hydropower and knowledge politics in the Puelwillimapu, Southern Chile. *Energy Res. Soc. Sci.* **2019**, *54*, 224–235. [CrossRef]
61. Ptak, T. Towards an ethnography of small hydropower in China: Rural electrification, socioeconomic development and furtive hydroscapes. *Energy Res. Soc. Sci.* **2019**, *48*, 116–130. [CrossRef]
62. Bradford, M.J.; Higgins, P.S.; Korman, J.; Sneep, J. Test of an environmental flow release in a British Columbia river: Does more water mean more fish? *Freshw. Biol.* **2011**, *56*, 2119–2134. [CrossRef]
63. Berkes, F.; Folke, C. *Linking Social and Ecological Systems: Management Practices and Social Mechanisms for Building Resilience*; Cambridge University Press: Cambridge, UK, 1998; ISBN 0-521-59140-6.
64. Olsson, P.; Folke, C.; Berkes, F. Adaptive comanagement for building resilience in social—ecological systems. *Environ. Manag.* **2004**, *34*, 75–90. [CrossRef] [PubMed]
65. Poff, N.L.; Allan, J.D.; Palmer, M.A.; Hart, D.D.; Richter, B.D.; Arthington, A.H.; Rogers, K.H.; Meyer, J.L.; Stanford, J.A. River flows and water wars: Emerging science for environmental decision making. *Front. Ecol. Environ.* **2003**, *1*, 298–306. [CrossRef]
66. Shirakawa, N.; Tamai, N. Use of economic measures for establishing environmental flow in upstream river basins. *Int. J. River Basin Manag.* **2003**, *1*, 15–19. [CrossRef]
67. Acreman, M. Linking science and decision-making: Features and experience from environmental river flow setting. *Environ. Model. Softw.* **2005**, *20*, 99–109. [CrossRef]

Article

Understanding the Complexity of Water Supply System Governance: A Proposal for a Methodological Framework

Fernando Gumeta-Gómez [1,*], Andrea Sáenz-Arroyo [1,2,*], Gustavo Hinojosa-Arango [3], Claudia Monzón-Alvarado [4], Maria Azahara Mesa-Jurado [5] and Dolores Molina-Rosales [6]

1 Departamento de Conservación de la Biodiversidad, El Colegio de la Frontera Sur, Unidad San Cristóbal de las Casas, Chiapas 29290, Mexico
2 Centro de Ciencias de la Complejidad (C3 UNAM), Universidad Nacional Autónoma de México, Ciudad de México 04510, Mexico
3 Cátedra CONACYT-CIIDIR Instituto Politécnico Nacional, Unidad Oaxaca, Oaxaca 71230, Mexico; ghinojosaar@conacyt.mx
4 Cátedra CONACYT-El Colegio de la Frontera Sur, Unidad Campeche, Campeche 24500, Mexico; cmonzon@ecosur.mx
5 Departamento de Ciencias de la Sustentabilidad, El Colegio de la Frontera Sur, Unidad Tabasco, Tabasco 86280, Mexico; mmesa@ecosur.mx
6 Departamento de Ciencias de la Sustentabilidad, El Colegio de la Frontera Sur, Unidad Campeche, Campeche 24500, Mexico; dmolina@ecosur.mx
* Correspondence: fernando.gumeta@estudianteposgrado.ecosur.mx (F.G.-G.); msaenz@ecosur.mx (A.S.-A.)

Abstract: The question of how the complexity of water governance may be understood beyond a heuristic concept remains unanswered. In this paper, we propose a Water Governance Complexity Framework to address the complexity of water governance. Through a literature review, rapid surveys, and 79 semi-structured interviews, we propose how this framework may be operationalized using different proxies and by applying it to the case of the water supply system for domestic use in Oaxaca, Mexico. In places such as the rural communities of Oaxaca, where the state plays a partially absent role in the water supply, we found legal pluralism and diverse formal and informal stakeholders in a multi-level structure. At the local level, four modes of governance were identified, resulting from seven institutional change trajectories. These trajectories result from linear (alignment) and non-linear (resistance and adaptation) interactions between local, state, and national institutions over different periods. We provide a pragmatic framework to understand complexity through the organization and historical configurations of water governance that may be applied globally, providing a necessary starting point and solid foundation for the creation of new water policies and law reforms or transitions to the polycentric governance model to ensure the human right to water and sanitation.

Keywords: institutional change; nestedness; governance mode; legal pluralism

Citation: Gumeta-Gómez, F.; Sáenz-Arroyo, A.; Hinojosa-Arango, G.; Monzón-Alvarado, C.; Mesa-Jurado, M.A.; Molina-Rosales, D. Understanding the Complexity of Water Supply System Governance: A Proposal for a Methodological Framework. *Water* **2021**, *13*, 2870. https://doi.org/10.3390/w13202870

Academic Editors: Athanasios Loukas and Luis Garrote

Received: 10 September 2021
Accepted: 11 October 2021
Published: 14 October 2021

Publisher's Note: MDPI stays neutral with regard to jurisdictional claims in published maps and institutional affiliations.

Copyright: © 2021 by the authors. Licensee MDPI, Basel, Switzerland. This article is an open access article distributed under the terms and conditions of the Creative Commons Attribution (CC BY) license (https://creativecommons.org/licenses/by/4.0/).

1. Introduction

Complexity analysis is an approach that is gaining strength when evaluating environmental policies [1]. In the water sector, complexity has been associated with problems such as environmental pollution, the overexploitation of aquifers, and the insufficient supply of adequate quality water to all people. This complexity results in the difficulty of fully understanding all variables that influence how these problems may be resolved [1,2]. These variables are linked to social, cultural, political, economic, technological, and environmental factors at different scales [3]. Moreover, complexity is an attribute assumed to be inherent in water governance, which is in part due to specific water-related problems [1,3,4] and the social–ecological system in which it is immersed [5–7]. Other studies have attributed complexity as a characteristic of adaptive [8] and polycentric governance models [9,10] due to their ability to incorporate uncertainty and feedback into the decision-making and water management process. Nonetheless, we argue that water governance can be complex not only as a characteristic of an adaptative or polycentric governance model but because of the complex problems it addresses across different jurisdictional, spatial, and temporal scales.

However, in the field of water governance, there is a lack of appropriate frameworks and pertinent variables to address complexity beyond a heuristic concept. In other research areas, such as the forest, fisheries, and aquaculture sectors, the diversity of stakeholders and institutions [11], the multilevel governance structure [12], legal pluralism [13], and the nestedness of the institution [14,15] have been proposed as approximations to understand the complexity of governance. However, these approaches largely ignore local processes, fail to identify multilevel structural elements or processes that contribute to complexity, or are unable to clearly define nestedness measures by overlooking the ambiguity between what is or what is not nested. In addition, we consider that approaching complexity via a single property is short-sighted while not being fully linked to the theory of complexity. In this theory, complexity attempts to holistically and synergistically understand the outcomes based on the interactions (e.g., exchange of information, goods, services, or energy) of system components (e.g., stakeholders and institutions), the evolution of the system, and the manner in which component interactions define the structure of the system while allowing for the emergence of qualities that cannot be either predicted or controlled [16]. In this sense, the question of how the complexity of water governance may be understood in a way that allows for analyses of empirical cases remains unanswered. Understanding the complexity of water governance is relevant due to the tendencies towards water management decentralization in many countries [17–19] and the existence of a multi-level process regarding the human right to water and sanitation that operates from global to local levels [20]. The decentralization of water management and the creation of new institutions and rights (e.g., human water rights) can create legal pluralism, resulting in new or different interactions between stakeholders and institutions. Evaluations of these new interactions will provide a solid foundation to establish new water policies, reform existing laws, or transition to more desirable polycentric or adaptative governance models [14] to ensure the human right to water and sanitation.

In this study, we propose a new conceptual–methodological framework called the Water Governance Complexity Framework, which is based on some elements of the Kooiman Interactive Governance Framework [14,21,22], to understand the complexity of water governance. To illustrate this framework, we used the water supply system for domestic use in Mexico. As in many other rural and suburban locations in Latin America, the inhabitants of Oaxaca use a variety of institutional arrangements to govern the water supply system. In this study, we argue that framing the governance of this system under the lens of complexity allows for its structure and function at state and national levels to be evaluated. For this, we aimed to answer three questions: (1) How is the current governance of the water supply system for domestic use in Oaxaca, Mexico, structured, and how does it function, considering the different jurisdictional levels? (2) Over time, how have institutional changes shaped the current governance structure of the water supply system for domestic use? (3) Can the water governance for domestic use in Mexico be considered complex?

2. Materials and Methods

2.1. Building the Water Governance Complexity Framework

To develop the Water Governance Complexity Framework, we began by establishing and linking basic concepts such as water governance and complexity. Subsequently, based on the Interactive Governance Framework (IGF) [14], we structured a new conceptual–methodological framework to understand the complexity of water governance.

We adopted a water governance definition in a broad sense to avoid controversy. Thus, we define water governance as a set of interactions used to make decisions among different stakeholders and institutions with common objectives to manage water resources [23]. These different stakeholders include governments, the private sector, and civil society [24]. Meanwhile, we distinguished institutions as formal rules, laws, and norms (e.g., constitutions, laws, regulations, and policies) and informal institutions as social agreements, as defined by North [25], which guide and regulate stakeholder decision making and actions. According to complexity theory, complexity is related to uncertainty and the challenges associated with predicting non-linear interactions between constantly changing entities [16]. A looser approach relates complexity to patterns and structures that are not easily describable or predictable [26]. If it is assumed that entities can be stakeholders and institutions and that the variables are their interactions (e.g., linear and non-linear) and change (e.g., institutional change), the first link is established between complexity and water governance. The second link between complexity and water governance is offered by the IGF proposed by Kooiman [14], as it considers the diversity of the governance system. A greater diversity of stakeholders and institutions may produce more dynamic and less predictable interactions between these entities. Ostrom sees the diversity of institutions as similar to that of ecological systems. In ecological systems, greater species diversity increases the structural complexity of biotic communities [27]. In ecological and economic systems, diversity helps to promote complexity and functionality [27,28]. However, diversity alone is not enough to produce complexity given that it requires that entities establish interactions and that new structures emerge as a result of those interactions [26]. These properties are vital in understanding the functionality of the system, which in our case, is water governance.

In this sense, the IGF offers a good starting point for integrating the variables of interaction, change, diversity, and complexity as properties of the stakeholders and institutions. The IGF is a relatively broad framework that addresses the *societal system*, defined as "the whole of interrelations among a given number of entities belonging to the natural and social worlds" [28]. According to the IGF, the societal system is made up of three parts that characterize it: the governing system (GS), system-to-be-governed (SG), and governing interactions (GI), in addition to the properties (i.e., *complexity*, *diversity*, *dynamics*, and scale), elements (i.e., image, instrument, and action), and orders (first- and second-order and meta-governance). The IGF mainly focuses on interactions to solve social problems and create opportunities, emphasizing interactions as its main innovation [15]. In this study, we focus on the properties of the GS in the first and second orders of governance. A full description of the other framework components can be found in Kooiman [14] and Kooiman and Bavinck [22].

Our proposal includes the following:

- The governing system encompasses the "total set of mechanisms and processes that are available for guidance, control, and steerage of the system-to-be-governed" [22].
- Properties are common concepts and measures that are used to understand the qualities of the system-to-be-governed and the governance system, such as superposition, links, interactions, and interdependencies [21]. The IGF considers diversity, dynamics, complexity, and scale as concepts, and measures commonalities.
    - Diversity is defined in terms of variation in the attributes or characteristics [26] of stakeholders and institutions [15] in the GS, SG, or GI [21]. Bavinck and Kooiman propose legal pluralism as a proxy for GS in the fisheries and aquaculture sectors [15].
    - Dynamics "create the potential for change" [15]. Bavinck and Kooiman propose institutional change as a proxy for GS in the fisheries and aquaculture sectors [15]. The principal analysis of institutional theory focuses on how stakeholders, institutions, and arrangements change over time [29]. The analysis also focuses on institutions that do not change or resist change due to stagnation, atrophy, or robustness [29].
    - Scale "represents the level at which the combined effects of diversity and dynamics can be best observed and analyzed" [21]. Following Gibson et al. [30], we clarify that scale and level are two different but related aspects. Scale refers to any dimension (e.g., spatial, temporal, and jurisdictional), and level refers to the unit of analysis in a different place on a given scale.

- Order of governance focuses on different processes.
    - First-order governance refers to the processes that deal with day-to-day problems. In this order, the stakeholders create opportunities each day [31] to solve operational problems related to supply, prices, costs, and user satisfaction. This first order of governance refers to what other authors consider to be management [32].
    - Second-order governance "focuses on the institutional arrangements within which first-order governance takes place" [21]. In this order, the institutional design and arrangement are expressed to allow, sustain, and focus governance [22]. Kooiman and Bavinck [15] consider a high-level expression of such institutional arrangements as the state, market, and civil society.

Our approach differs from the Kooiman IGF by viewing complexity as an umbrella property encompassing diversity and dynamics. Additionally, we propose incorporating nestedness as a property (Figure 1). Nestedness is a property linked to the interactions between entities (e.g., stakeholders and institutions). The importance of this property lies in analyzing the influence of the structure of the system on the behavior of the subsystem [33]. In this sense, emphasis is placed on the nestedness of the scalar property [34], which for our purposes represents a jurisdictional scalar (i.e., local, state, or national levels). We argue that considering complexity as a supra-property can help reconcile the IGF approach with the conceptualizations derived from complexity theory. Therefore, we consider scale to be a cross-sectional condition of all properties, as it is not practical to begin an analysis without clearly defining the scales or levels under observation [35].

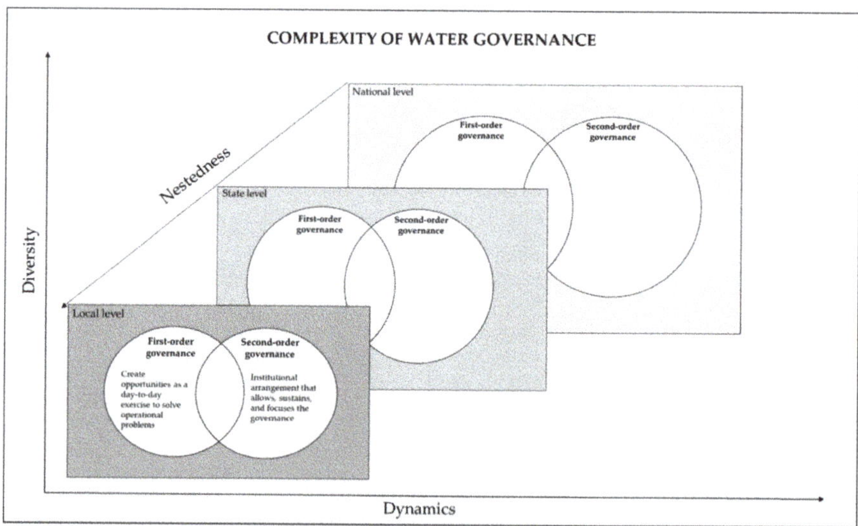

Figure 1. The Water Governance Complexity Framework proposes the analysis of diversity, nestedness, and dynamics properties in at least three levels in jurisdictional (i.e., local, state, and national) and temporal (levels are defined a posteriori according to periods of institutional change) scales in the first and second orders of governance. Source: Adapted from Kooiman and Bavinck [22].

2.2. *Case Study: Water Supply System for Domestic Use in Oaxaca and Mexico*

This study applies the Water Governance Complexity Framework to analyze the complexity of the governance of the water supply system for domestic use in Oaxaca and Mexico. This system refers to water obtained from freshwater resources using hydraulic infrastructure, which allows for its storage, treatment (to ensure it is suitable for human consumption), and transport (to satisfy the food, health, and hygiene needs of each household). Our analysis focuses on formal and informal stakeholders and institutions immersed or involved in managing the water supply system for domestic use to solve appropriation and provision problems [36] through jurisdictional and temporal scales.

At the national level, we reviewed the institutions, stakeholders, and institutional changes of the nineteenth, twentieth, and twenty-first centuries related to the management of the water supply system for domestic use. At the state level, we selected Oaxaca and its legislation related to water for domestic use. We chose 13 rural communities in the Mixtecan Alta region in Oaxaca (Figure 2) to explore the diversity of stakeholders and institutions, nestedness, and dynamics at the local level. In its broadest sense, we emphasize that community refers to a social unit that shares things in common, such as norms, religion, values, or identity [37]. We selected Oaxaca because the local government systems are considered unique and relatively more autonomous than those of other Mexican states [38]. The 13 rural communities selected for this study (Figure 2) are indigenous and cover different political and administrative configurations. Six communities are municipal seats (San Francisco Teopán, Santa Magdalena Jicotlan, Concepción Buenavista, Santiago Ihuitlán Plumas, San Juan de los Cues, and Santiago Tepetlapa), and seven communities are municipal agencies (El Enebro, San Miguel Aztatla, Santa Cruz Corunda, San Antonio Abad, La Mexicana, Santiago Quiotepec, and Santa Cruz Capulalpam). Municipal agencies are subdivisions of the same municipality that encompass peripheral population centers and are subordinate to the municipal seat. In the municipal seat, the municipal council is established and acts as the leading local authority in the municipality.

2.3. *Operationalization, Data Collection, and Analysis*

The diversity, nestedness, and dynamics of the Water Governance Complexity Framework were approached by assessing proxies for legal pluralism, formal and informal stakeholders, nestedness among jurisdictional levels, and institutional change at national and local levels, following the proposal of Kooiman [14,15,39]. We present the proxies, their operative definitions (Table 1), the methodology used to obtain data, and the implemented analyses in detail in the following subsections.

2.3.1. Diversity

This study addresses diversity through legal pluralism in managing the water supply system for domestic use. According to Tamanaha [17], a "simple" definition of legal pluralism considers the role of social actors when identifying more than one source of "law" (institutions) or normative order within a social arena. Sources of normative ordering include official legal systems (formal institutions); customary, cultural, religious, economic, functional, and community normative systems (general informal institutions according to North [25]); or even multiple legal systems, both formal and informal. According to the IGF, the result is the existence of multiple legal systems (institutions) that determine the governing system and influence the governance object [15].

To address legal pluralism and formal and informal stakeholders, we implemented a multi-level approach, which first evaluated the official legal system at national and state levels for different formal institutions that could potentially overlap or align in the management of the water supply system for domestic use. We first reviewed the Ley de Aguas Nacionales (National Water Law; LAN, acronym in Spanish). Likewise, we reviewed other laws and regulations that could influence this system. First, a search was carried out in the Political Constitution of the United Mexican States of 1917 (National Constitution) and the Political Constitution of the Free and Sovereign State of Oaxaca (State Constitution) using Nitro PDF Pro v. 12.4.0.25.9 with the Spanish keywords "agua potable", "agua para consumo humano", and "agua para uso domestico". From this search, we identified articles directly related to the management of the water supply system for domestic use. This national and state constitutional review allowed for the identification of other laws at these levels, such as:

National level

- Agrarian Law (regulates land tenure and the collective rights of the 13 selected communities);
- The General Law of Ecological Balance and Environmental Protection (LGEEPA).

State-level

- The State of Oaxaca Law for Potable Water and Sewerage;
- The State of Oaxaca Law for the Rights of Peoples and Indigenous Communities.

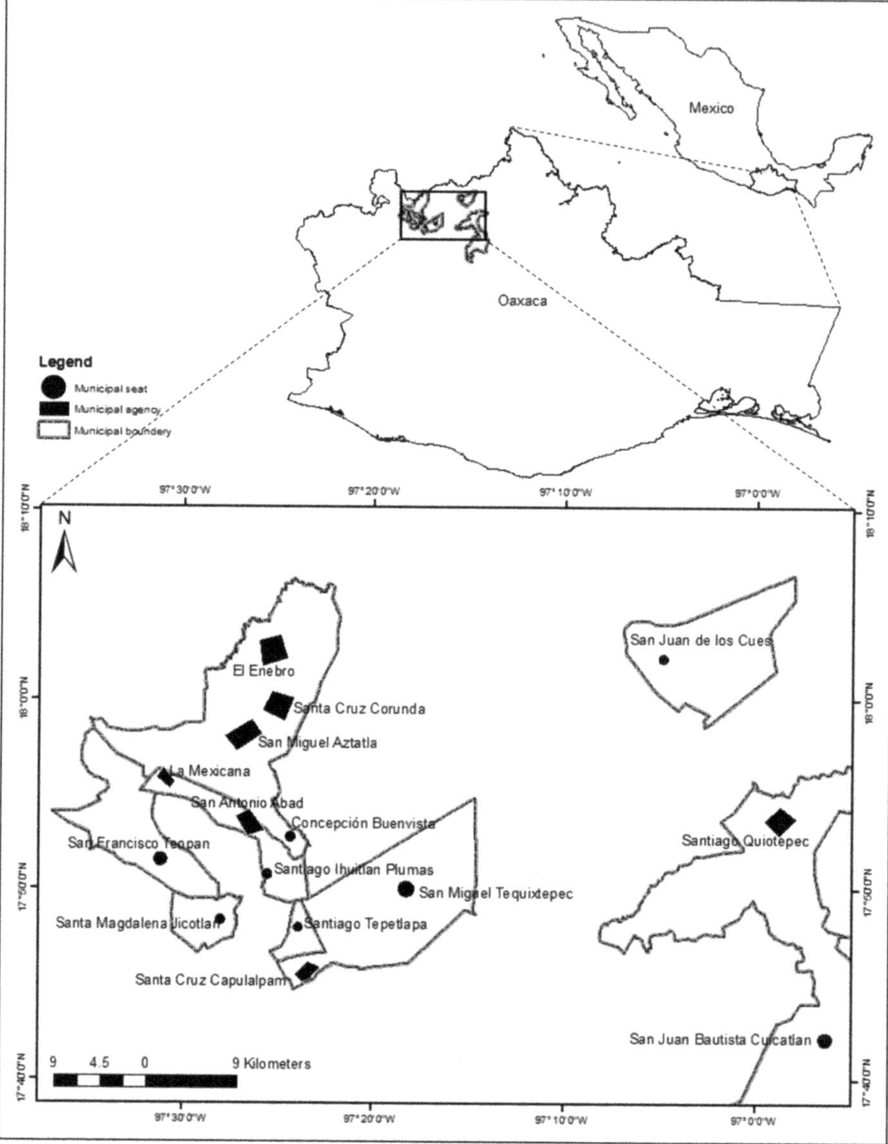

Figure 2. Macro- and micro-locations of the 13 rural communities in Oaxaca, Mexico, selected for this study. The map shows the political and administrative orders of the municipalities (dotted line), municipal seats (circles), and municipal agencies (rectangles). Municipalities are named after their municipal seats. Source: Prepared by the authors from governmental vector data.

Table 1. Operationalization of the properties that comprise the complexity of the governance of the water supply system for domestic use through the diversity, nestedness, and dynamics of the Interactive Governance Framework (IGF). Sources: Prepared by the authors based on Bavinck and Kooiman [15].

Properties	Proxies	Description
Diversity	Legal pluralism	Different formal and informal institutions (laws or regulations) that intervene in the right to administer, manage, or regulate the water supply system for domestic use
	Formal stakeholders	Stakeholders recognized by different formal institutions
	Informal stakeholders	Stakeholders not recognized by different formal institutions
Nestedness	Nestedness of formal and informal stakeholders and institutions	Interactions between different stakeholders belonging to different jurisdictional levels (municipal agency, municipality, state, and nation) in 10 different activities of the first and second orders of governance
Dynamics	Institutional change at the national level	Changes in stakeholders and institutions related to water management for domestic use at the national level during the nineteenth, twentieth, and twenty-first centuries
	Institutional change at the local level	Changes in stakeholders and institutions at the community level during the nineteenth, twentieth, and twenty-first centuries

Subsequently, we investigated the informal institutions and stakeholders involved in managing the water supply system in the 13 rural communities. A rapid survey was administered to local authorities in 2019 and consisted of four questions classified according to whether the activities corresponded to first-order (operational) or second-order governance (Table S1, Supplementary Materials).

Finally, we cross-referenced the results obtained from formal and informal institutions and stakeholders to define the structure influencing the governance of the water supply system for domestic use in Oaxaca, Mexico.

2.3.2. Nestedness

A semi-structured interview was implemented with key stakeholders within the 13 communities to obtain data on cross-level interactions among stakeholders of different jurisdictional levels. Key stakeholders included two main groups: (1) current or past hydraulic network operators and (2) officials from the municipal council or municipal agency. These stakeholder groups were identified from the rapid survey. We identified the key stakeholders following the snowball method, which identified potential interviewees and then asked them for recommendations on whom to interview later [40]. The semi-structured interview contained a matrix in which the rows represented the ten activities belonging to the first and second governance orders (Table S2, Supplementary Materials). The columns represented the stakeholders directly or indirectly responsible for the water supply system for domestic use in the studied communities. We obtained a total of 79 semi-structured interviews (La Mexicana (4), Santa Cruz Capulalpam (4), San Francisco Teopan (5), El Enebro (7), San Antonio Abad (3), Santa Cruz Corunda (3), San Miguel Aztatla (7), Santiago Quiotepec (9), Santa Magdalena Jicotlán (10), Concepción

Buenavista (12), Santiago Ihuitlan Plumas (9), San Juan de los Cues (2), and Santiago Tepetlapa (4)).

Each stakeholder was characterized according to their respective jurisdictional level (Table S3, Supplementary Materials). Subsequently, the obtained matrix was analyzed in two ways. First, we conducted a descriptive analysis of the cross-level interactions reported by the interviewees from each community. Second, we implemented a metric analysis using the Nestedness based on the Overlap and Decreasing Fill (NODF) methodology proposed by Almeida-Neto et al. [41]. Recently, NODF has been used to analyze social and commercial networks [34]. According to Almeida-Neto et al. [41], NODF is based on two simple properties, decreasing fill (DF) and paired superposition, to calculate the entire nestedness of a binary matrix.

For this reason, the first matrix obtained in the first step with binomial presence (1) and absence (0) data (Table S3, Supplementary Materials) was split into two groups: one matrix for municipal seats (Table S4, Supplementary Materials) and another for municipal agencies (Table S5, Supplementary Materials). As mentioned earlier, municipal agencies are hierarchically subordinate to the municipal seat and should hypothetically be nested. The NODF analysis was carried out with the open-source online program NeD (Nestedness for Dummies) of the Joint Research Center (http://ecosoft.alwaysdata.net/ accessed on 15 February 2021) created by Strona et al. [42]. The NeD program provides information such as the nestedness index and the probability levels after comparing the matrix under evaluation with a certain number of null matrices. According to Ulrich and Gotelli [43], the null matrices can be obtained through five different null models: EE (equiprobable row totals and equiprobable column totals), CE (proportional row totals and proportional column totals), FE (fixed row totals and equiprobable column totals), EF (equiprobable row totals and fixed column totals), and FF (fixed row and fixed column totals). The null model chosen to test nesting significance is decisive with regard to the results obtained [42].

Finally, we compared the results obtained from both approximations to generate a complete analysis of nestedness and the advantages of each approximation. A descriptive approach allowed us to obtain an overview of the results without losing detail. For its part, an approximation based on a metric NODF can help shed light on whether it is nested and to what degree it is nested, decreasing the ambiguity of the descriptive approach.

2.3.3. Dynamics

We carried out a literature review of books and scientific papers to analyze the trajectory of changes to laws, norms, and regulations or reforms such as those of the Mexican Water Law [44–50]. We also reviewed institutional changes in other laws such as those of the Agrarian Law [51]. For the analysis of institutional change at the local level, we applied three open questions in a semi-structured interview to collect retrospective information on the trajectories of the stakeholders and institutions responsible for managing the water supply system for domestic use in the 13 communities included in this study (Table S6, Supplementary Materials). For this section, we interviewed older people and recognized experts in each community who either held important positions or had experience managing water for domestic use. In the end, we compared the information obtained from the institutional changes at the national level with the institutional changes that were documented at the local level in the 13 communities studied.

The method we are proposing is summarized in Figure 3.

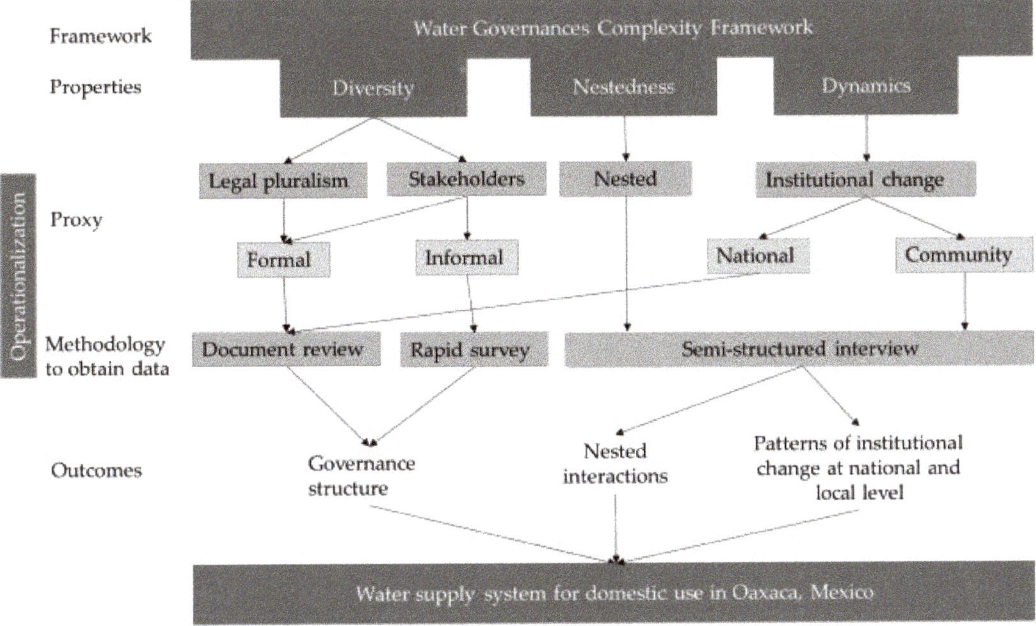

Figure 3. Scheme of the operationalization of the Water Governance Complexity Framework, the proxies used to address them, the methodology implemented to obtain data, and expected outcomes, using the case of the water supply system for domestic use in Oaxaca, Mexico.

3. Results

We present the results obtained from the proxies of the diversity, nestedness, and dynamics properties in different sections. In the last paragraph of each section, we present a cross-analysis of the different focuses and analytical approaches with which diversity, nestedness, and dynamics were evaluated. We conclude the results section with a cross-analysis of the results obtained from the different proxies of each property, integrating them through the Water Governance Complexity Framework.

3.1. Diversity

We identified multiple operating institutions or legal systems that overlapped or aligned at the local level. Plural institutions are part of the national-state legal system, as they are derived from the "supreme law" of the National Constitution, which agree with those of the Oaxaca State Constitution.

Article 27 establishes that water ownership pertains to the nation and that the nation has the right to transmit its property to individuals [52]. By declaring itself as the legitimate owner, the federal executive branch possesses all the regulating property rights of the water supply system for domestic use. However, it also establishes that:

- Landowners can extract subsoil waters and take advantage of natural outcrops within one plot. They are granted the right to access and use the water and the property right to exclude other individuals from accessing that water. However, they cannot provide water services to other individuals or populations, and landowners must first give concessions.
- Population centers that communally operate can use the water that belongs to or has been returned to the community. These centers have the right to access, use, and manage water to meet the needs of their populations and retain the right to exclude other communities from accessing their water.

The preceding statements are reaffirmed and specified in three laws derived from Article 27: The LAN of 1992 [53], the Agrarian Law of 1992 [54], and Ley General del Equilibrio Ecologico y Proteccion al Ambiente (General Law of Ecological Balance and Protection of the Environment; LGEEPA; acronym in Spanish) of 1998 [55]. The LAN is the sole law that establishes a multi-level structure for water governance. At the national level, the National Water Commission (CONAGUA; acronym in Spanish) is the autonomous and decentralized body of the Ministry of Natural Resources and the Environment (SEMARNAT; acronym in Spanish). CONAGUA is responsible for the administration, regulation, and consultation of water management in Mexico [53,56]. The LAN establishes the Watershed Council to manage Hydrological–Administrative Regions (RHAs; acronym in Spanish) at the state level.

The RHAs include groups of basins and municipal territories to facilitate the administration and integration of socioeconomic data [57]. The Watershed Council is meant to provide support and advice among CONAGUA; municipal, state, and national governments; user representatives; and civil society organizations [53]. At the local level, CONAGUA recognizes and grants access, use, and management rights to the state, municipality, non-governmental organizations (NGOs), and the private sector. Nonetheless, as the state and municipalities are considered subdivisions of the Nation-State, they do not have the right to exclude any individuals due to the recent reform to constitutional Article 4, which establishes the human right to water. In cases involving private companies and NGOs, the right of exclusion is upheld. In addition, the Agrarian Law and the LGEEPA reaffirm the rights of agrarian communities (e.g., ejidos and Bienes Comunales; article 52 of the Agrarian Law) to own water for common use for both agricultural and domestic purposes [54]. The LGEEPA also recognizes indigenous communities, which are not necessarily considered within agrarian communities. Article 15 (section XIII) of the LGEEPA establishes that the Nation-State must guarantee the rights of indigenous peoples regarding the sustainable use and exploitation of natural resources, which implicitly includes water for domestic use [55]. Although these laws are linked to Article 27, they seem to address other non-municipal social contexts, unlike the LAN. However, in the case of municipalities that are also indigenous or that have agrarian communities, these laws overlap.

Article 115 explicitly designates municipalities as responsible for the management of the water supply system for domestic use at the local level, establishing how this responsibility should be carried out in coordination with CONAGUA and the Watershed Council with regard to the planning, execution, administration, and management of national water resources [58]. This article matches those established with the LAN.

On the other hand, Article 40 of the National Constitution establishes that every Mexican state can create its constitution [52], including establishing other laws designed to regulate and manage water for domestic use. In Oaxaca, the State Law of Potable Water and Sewerage [59] establishes new stakeholders at local levels. The State Water Commission is responsible for developing the water supply system for domestic use at the state level. Article 17 recognizes municipalities and citizen water committees as stakeholders at the local level if no municipal operations agency is present. The water committee can promote the construction, conservation, maintenance, rehabilitation, and operation of its water, piped water, and sewer systems [59]. In this case, state water law in municipalities aligns with Article 115 and the LAN. However, with water committees, both national laws overlap with state law.

Article 2 of the National Constitution stipulates that each state is responsible for formulating and promoting its laws regarding the rights of indigenous peoples and communities. In the case of Oaxaca, the law of the Rights of Indigenous Peoples and Communities [59] recognizes their social, cultural, religious, political, and self-determination rights. In this sense, the self-determination rights of indigenous community stakeholders that manage water for domestic use are recognized. This state law matches with the LGEEPA and Agrarian Law but overlaps with the LAN and Article 115 of the National Constitution in indigenous municipalities. It should be mentioned that because indigenous communities have the right to self-determination, in addition to the property rights to access, use, and manage water, the right to exclusion may also be included (e.g., if indigenous institutions consider suspending the water service as a sanction for any fault).

In addition, we identified two informal stakeholders not established by the existing national and state institutions through the surveys conducted in the 13 communities selected for this study (Table 2): the municipal agent (in 53% of the studied communities) and the assembly of water users (in 76.9% of the studied communities).

Table 2. Stakeholders identified by the survey administered in the 13 rural communities. The questions correspond to the first (operability) and second (institutional arrangement) orders of the Interactive Governance Framework (IGF) [14]. NP = no payment for water services, MA = municipal agent, WC = water committee, MC = municipal council, WUA = water users assembly.

Communities	Responsible for Managing the Water Supply System for Domestic Use	Decision Makers for Domestic Water Issues	Recipients of Payments for the Domestic Water Service	Decision Makers for the Money Collected from Payments to the Water Supply Service
La Mexicana	MA	WUA	NP	NP
Santa Cruz Capulalpam	MA	WUA	NP	NP
San Francisco Teopán	WC	WUA	NP	NP
El Enebro	MA	WUA	NP	WUA
San Antonio Abad	WC	WUA	WC	WUA
Santa Cruz Corunda	MA	WUA	WC	MA
San Miguel Aztatla	WC	WUA	WC	WC
Santiago Quiotepec	WC	WUA	WC	WC
Santa Magdalena Jicotlán	MC	WUA	MC	MC
Concepción Buenavista	MC	WUA	MC	MC
Santiago Ihuitlan Plumas	MC	MC	MC	MC
San Juan de los Cues	MC	MC	MC	MC
Santiago Tepetlapa	MC	MC	MC	MC

The information obtained from these formal national and state institutions with local impacts was complemented with information obtained in the field regarding informal stakeholders. This information was used to identify the multi-level structure of the institutions and stakeholders immersed in water governance in Oaxaca and Mexico (Figure 4).

3.2. Nestedness

Four municipal seats reported cross-level interactions with stakeholders from the two highest jurisdictional levels of the state (except Santiago Tepetlapa) and nation. Conversely, among eight municipal agencies in which the municipal agent or water committee was responsible for the water supply system for domestic use, San Antonio Abad, Santa Cruz Corunda, and El Enebro did not report interactions with any stakeholder at higher jurisdictional levels (i.e., the state level), nor did San Antonio Abad with stakeholders at the national level (Figure 5b). Reports of interactions between San Antonio Abad, San Miguel Aztatla, and Santiago Quiotepec with their municipalities (e.g., Santiago Ihuitlan Plumas, Concepción Buenavista, and San Juan Bautista Cuicatlan, respectively) were low (21–40%). A similar situation was present in the interactions reported between La Mexicana and Santa Cruz Corunda with stakeholders at the national jurisdictional level (21–40%), such as CONAGUA. The four municipal seats mainly presented interactions with the national jurisdictional level, which was recognized through interviews (81–100%, Figure 5a). In the ten activities analyzed, it should be noted that the 13 communities in this study reported interactions with the assembly of water users with regard to decision making. Likewise, the commissariat of communal and ejidal assets (agrarian council) as well as migrant users, either individually or in an organized manner (directive), were reported as stakeholders involved in the first order of governance of the water supply system for domestic use in all communities (Table 3).

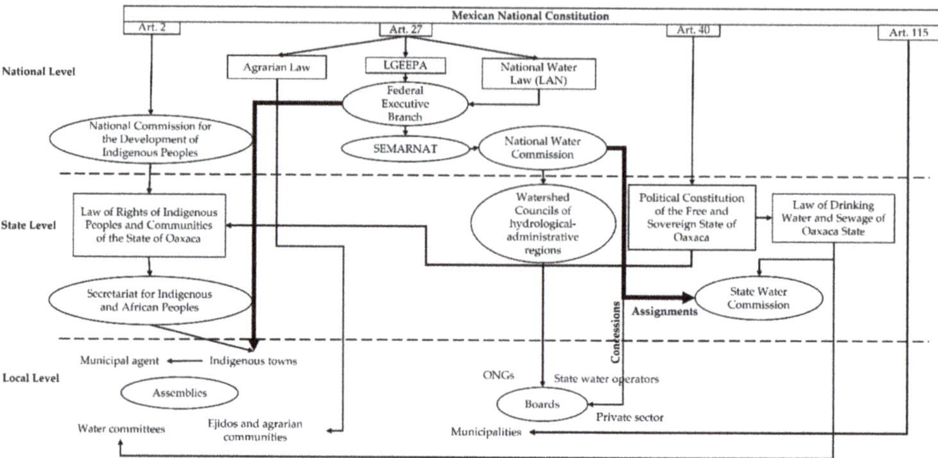

Figure 4. The multi-level structure of the governance of the water supply system for domestic use in Mexico. The squares and circles refer to the laws (institutions) and stakeholders (circles), respectively. Arrow thickness only serves to differentiate among arrows when they intersect. LGEEPA: Ley General del Equilibrio Ecologico y la Protección al Ambiente. SEMARNAT: Secretaria del Medio Ambiente y Recursos Naturales. Adapted from: Gumeta-Gómez et al. [60].

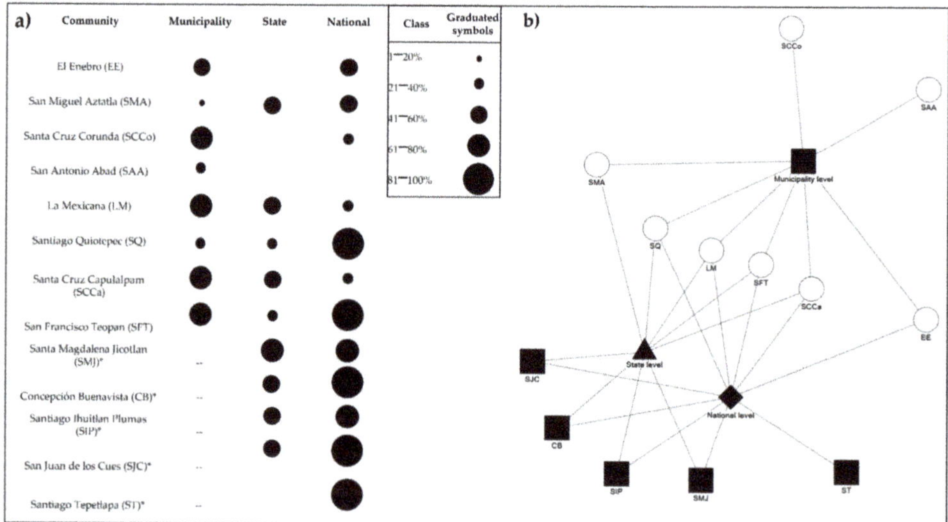

Figure 5. (**a**) Percentage of interviewees from each community who mentioned some interaction with the different stakeholders belonging to higher jurisdictional levels. Communities marked with an asterisk (*) are municipal seats that cannot be nested within themselves, so interactions at the municipal level were not considered. (**b**) Interaction scheme of the 13 communities studied with other actors at higher jurisdictional levels: municipal (square), state (triangle), and national (diamond). The communities where a municipal agent or a water committee administers the water supply system for domestic use are shown in a circle and the municipalities in squares. Abbreviation definitions can be found in (**a**).

Table 3. Interactions between actors at the local level within the 13 studied communities. The commissariat of communal or ejidal assets corresponds to those responsible for the ejido agrarian territory or agrarian community with collective land tenure. * Stakeholder responsible for the operation of the water supply system for domestic use in the community.

Community Level	Municipality	Municipal Agent	Water Committee	Commissariat of Communal Assets	Commissariat of Ejidal Assets	Assembly of Water Users	Migrant Water Users	Migrant Water Users Directive	Neighboring Municipality	Tourism Committee
La Mexicana	x	*		x		x		x		
Santa Cruz Capulalpam	x	*		x		x	x			
San Francisco Teopan	x		*	x		x	x			
El Enebro	x	*		x	x	x		x		
San Antonio Abad	x	x	*	x		x	x			
Santa Cruz Corunda	x	*		x	x	x	x			
San Miguel Aztatla	x	x	*	x	x	x		x	x	
Santiago Quiotepec	x	x	*	x	x	x	x			x
Santa Magdalena Jicotlan	*			x		x	x			
Concepción Buenavista	*			x		x		x		
Santiago Ihuitlan Plumas	*			x		x	x			
San Juan de los Cues	*	x		x		x				
Santiago Tepetlapa	*			x	x	x		x		

We did not observe significant nestedness in the results of the NeD analysis ($Z = 0.538$, $p > 0.05$) in any of the null models for the communities in which water committees were present or in which the municipal agent was responsible for the water supply system for domestic use (Table 4). However, in communities that are municipal seats, we obtained a significant nestedness value of 66.66 ($p > 0.001$) with the state and national levels. This result was consistent in all the null models (EE, CE, FE, FF, and EF). In all analyses of both community matrices, we used 50 random null matrices when calculating the Z value.

All municipal seats present cross-level interactions at the national level and to a lesser extent at the state level, which could explain the significant nestedness found with the NODF metric. In the case of the municipal agencies, half of them did not report cross-level interactions or only reported cross-level interactions with a single level (e.g., state or national), which could explain the non-nestedness of the group. However, we may consider that Santiago Quiotepec, La Mexicana, San Francisco Teopan, and Santa Cruz Capulalpam are nested or at least show a degree of nestedness, as they present cross-level interactions with the state and national jurisdictional levels.

Table 4. Nestedness results obtained with the Nestedness based on Overlap and Decreasing Fill (NODF) algorithm in the Nestedness for Dummies (NeD) online software for two groups: (1) communities in which water committees and municipal agents are responsible for supplying the water for domestic use and (2) municipal seats in which the municipality is responsible.

	Metrics	NODF_Total	NODF_Fill	NODF_Col
Water committee/municipal agent [1]	Index	64.516	60.714	83.33
	Z-Score	0.906	0.927	0.538
	RN	0.095	0.066	0.203
	Nested?	No [3]	No [3]	No [3]
Municipalities [2]	Index	46.154	40	66.66
	Z-Score	10825960642	NA (std = 0)	11728124031
	RN	0	0	0
	Nested?	Yes	NA (std = 0)	Yes [4]

[1] Communities with a water committee or municipal agent: Enebro, San Miguel Aztatla, Santa Cruz Corunda, San Antonio Abad, Santa Cruz Capulalpam, Santiago Quiotepec, La Mexicana, and San Francisco Teopan. [2] Municipalities: Santa Magdalena Jicotlan, Concepción Buena Vista, Santiago Ihiutlan Plumas, San Juan de Los Cues, and Santiago Tepetlapa. [3] $p > 0.05$. [4] $p > 0.001$. NODF_FILL: Nestedness of the fill. NODF COL: Nestedness of the column. NA: Not applicable.

3.3. *Dynamics*

We identified institutional changes in the Mexican water sector that determined the prevalence of one stakeholder over another in different periods, the creation or emergence of new institutional arrangements, or the formalization of existing stakeholders (recognition in written laws) in post-revolutionary Mexico (nineteenth, twentieth, and twenty-first centuries). These institutional changes in water governance in Mexico were framed in three periods: (1) pre-centralization, (2) centralization, and (3) decentralization [45–47] (Figure 5).

During the pre-centralization period at the beginning of the nineteenth century, colonial heritage prevailed, and water governance was considered a local matter to be handled between municipal governments, state governments, and individuals. The Mercedes (i.e., sanctioned use over a stream or spring), ordinances (i.e., the distribution of water to citizens by judges), and repartimiento (i.e., legal framework of the Repartimiento de Aguas established by the Spanish Crown in 1560) were recognized in the first Constitution of 1857. Article 27 of the National Constitution of 1857 guaranteed that the Mexican nation had to preserve property rights, including those over the water in rivers and springs [61].

The first law that gave the Federal Executive Branch control over rivers, canals, and navigable water bodies was the Ley de Vias Generales de Comunicación of 1888 [46]. However, its role regarding the ownership of national waters remained ambiguous [61]. For Rolland and Vega [48], the centralization process began with the first Ley de Aprovechamiento de Aguas de Jurisdicción Federal in 1910. This law established the Federal Executive Branch as the sole owner of all national waters. Centralization could be associated with the economic, social, and political power of controlling the water [60] and with the new technologies related to water use, health and hygiene, and distribution [62]. According to Escobar [50], municipal councils and states began to lose control over their waters with this law, as power was concentrated within a single national stakeholder. With the National Constitution of 1917, changes in water governance were introduced in article 27. These changes gave the federal government the power to issue laws regulating national waters and collect a tax for their concessions [61]. The creation of the Secretariat of Hydraulic Resources (SHR) in 1948 and the issuance of the Regulation of the Federal Drinking Water Boards in 1949 executed the transfer of municipal or state control of the water supply system for domestic use to the federal government through the Federal Water Boards (FWB) in the case of large cities, or Local Water Boards (LWB) in municipalities [62]. An LWB was made up of a municipal council member, town users, and a state government representative who reported to the federal government [63]. During federalization, most hydraulic infrastructure investments were made to bring water from sources (e.g., springs, wells, rivers, and lakes) to homes and were implemented by the now extinct Secretary of Hydraulic Resources [64]. The decentralization process began in 1980, when the federal government handed all drinking water and sewage systems that it managed and operated through the FWB and LWB over to state or municipal governments [63]. The federal government intended to partly correct the regional development imbalances caused during the centralization period [47], retaining the role of establishing regulations and the right of alienation by controlling water concessions [48]. The states created different operating bodies of the water supply system for domestic use within municipalities [62]. In many cases, control was strictly passed on to the municipality, and in others, the state maintained a water board model, creating State Water Boards (SWB) with broad collaboration from the municipal councils [63]. With the reform of article 115 in 1983, the responsibility of the water supply system was transferred solely to the municipalities [47]. Later, the Water Law reform of 1992 that established the decentralization process was reaffirmed by The North American Free Trade Agreement (NAFTA) in 1994. This change allowed municipal authorities to grant licenses to private companies to supply water for domestic use. These licenses were granted in Mexico City, Cancun, Navojoa, Aguascalientes, and Puebla [45,65].

Parallel to the institutional changes regarding water at the national level, other national laws or reforms were also created by the end of the twentieth century, including LGEEPA in 1986 and constitutional reforms to Article 2 that recognized the inalienable rights of indigenous peoples in 1992, such as rights to natural resources (e.g., water) within their lands.

At the local level, 70 interviewees did not remember or mention institutional changes regarding the management of the water supply system for domestic use in their communities. Notably, in the communities of La Mexicana, Santa Cruz Capulalpam, San Francisco Teopan, El Enebro, San Antonio Abad, Santa Cruz Corunda, San Miguel Aztatla, and Santiago Quiotepec, the interviewees mentioned that "it has always been like this," indicating that the stakeholder currently managing the water had done so for as long as they could remember. For example, a 55-year-old municipal agent of San Miguel Aztatla said that "the water committee has been working for more than 100 years in our community [...], according to their uses and customs" (i.e., indigenous institution).

Only nine interviewees provided relevant information that could be used to trace the trajectories of institutional changes within their communities, neighboring communities, and the Mixtecan Alta region. The nine interviewees were between the ages of 37 and 85 years old (Table 5). Based on their responses, we can identify the presence of the Papaloapan Commission (PC) between 1954 and 1979 in the Mixteca Alta region. Another stakeholder is the State Water Board (SWB) that controls the provisioning system in Concepción Buenavista, Santiago Ihuitlan Plumas, and other municipalities such as Santa Magdalena Jicotlan. In the case of Concepción Buenavista, a transition occurred from a water committee in 1983 to an SWB in 1985. The last recent change was from the SWB to fully municipal management in 1999. In the case of Santiago Tepetlapa and San Juan de los Cues, a change from a water committee to municipal management was reported in the last decade, which seems to have been motivated by endogenous issues in the communities, such as high rates of migration and the ability of the municipality to request funds for hydraulic works.

Table 5. Local knowledge of institutional changes in the management of the water system for domestic use with implications in five communities of the Mixteca Alta region in Oaxaca, Mexico.

Interviewee Age and Residence	Quote
A 79-year-old interviewee from Concepción Buenavista	"Before there was a water committee . . . it is no longer done like that . . . now it is the municipality, and they only report to the federal and state governments."
A 64-year-old interviewee from Concepción Buenavista	"The water committee that existed, if I remember correctly, as in '83 (1983). Later it became the Potable Water Board from '85 or so . . . managed by the Coordinator of Water Works Systems of Oaxaca. However, they wanted to put water meters on us, which did not suit us, and the people thought that if the municipality could take charge of it . . . that was like in '99. The coordinator took charge of several municipalities in the region like Santiago Ihuitlan Plumas, Tepelmem Villa de Morelos, Santa Magdalena Jicotlan, and many others."
An 85-year-old interviewee from Concepción Buenavista	"Years ago, the committee disappeared because there are not many people . . . the Commission of Papaloapan trained us, and we managed hydraulic works for the community. He helped us get the concession of the well, too. Before, the school also used to count on the committee (for water issues) . . . they were supported by pure money from the town. In '85, the first network was made; the committee checked the proper use of water, there was a committee regulation . . . then it passed to the municipality".
A 37-year-old interviewee from Concepción Buenavista	"The Commission of the Papaloapan helped us build the hydraulic water network . . . helped us train us to use it. First, the Papaloapan commission was in charge . . . I think it was on the part of the state; then they left it to the municipality."
A 48-year-old interviewee from Concepción Buenavista	"Now, the Councilor of finance (part of the municipal council) is in charge of the drinking water system (water for domestic use) . . . , before 25–30 years . . . there was a water committee; it was left due to the failures of people (the managers assigned as part of the committee of water)."
An 85-year-old interviewee from Santiago Ihuitlan Plumas	"The people worked so that (water) would not be lacking . . . the water service began in 1973. In 1954, the Papaloapan (the commission) helped . . . making "pretiles" (stone borders) to retain the water and soil . . . the hills were going . . . , the land, until the Papaloapan. The Papaloapan with authority (municipal council) managed the water (hydraulic system) . . . then the Papaloapan (the commission) left, and only the municipality remained (administering the water system for domestic use)".
An 84-year-old interviewee from Santiago Ihuitlan Plumas	"The commission of the Papaloapan helped us with the hydraulic work . . . , gave us the money and taught us how to do it. The stone borders ("petriles") helped us with the commission of the Papaloapan. In 1974, the Papaloapan Commission was withdrawn. There was a water committee . . . the '70s and '80s . . . they did not feel like it".
A 55-year-old interviewee from Santiago Tepetlapa	"We had (water committee) . . . 10–12 years ago, the water committee work. The water committee disappears because of a lack of people to provide service (a position occupied as a service to the community and free of charge for a specified period). There are almost no people in the town . . . the older people are left alone. We are very few men (young adults)."
A 38-year-old interviewee from San Juan de los Cues	"Before, about ten years ago, there was a water committee . . . , but the town decided that we would administer the drinking water (it refers to the municipal council to which it belongs). I believe that the people left it to us (the water supply system for domestic use) . . . because we could get works (hydraulic works). It is necessary to rehabilitate the dam and wells and build new wells to solve the drought problems that the town suffers."

By comparing the information on institutional change at the national and local levels, we identified that periods of institutional change at the national level (pre-centralized, centralized, and decentralized) did not permeate in all communities at the local level, especially in communities where the municipal agent or a water committee was responsible for the water supply system. In the case of municipalities, institutional changes at the national level permeated differently. For example, in Santiago Ihuitlan Plumas and possibly Santa Magdalena Jicotlan, institutional change coincided with change at the national level, where power was centralized through the LWB (with the Papaloapan Commission as the federal representative) only to be later decentralized to the municipality. In the case of Concepción Buenavista, decentralization occurred in two phases. In the first phase, decentralization resulted in the responsibility of the water supply systems for domestic use to be passed to the state through the SWB. In the second phase, control passed entirely to the municipalities. Recent institutional changes regarding the rights of indigenous peoples and the environment at the national level have not had notable impacts to date on any institutional change related to water management in the communities. Instead, these institutional changes at the national level have been made to formalize indigenous institutions (municipal agent and water user assembly).

3.4. *Intertwining Properties to Address the Complexity*

Using the Water Governance Complexity Framework, we cross-analyzed the results obtained from the different proxies of the properties of diversity, nestedness, and dynamics in the three jurisdictional levels (national, state, and local) and the four post-revolutionary periods of institutional change (Figure 6).

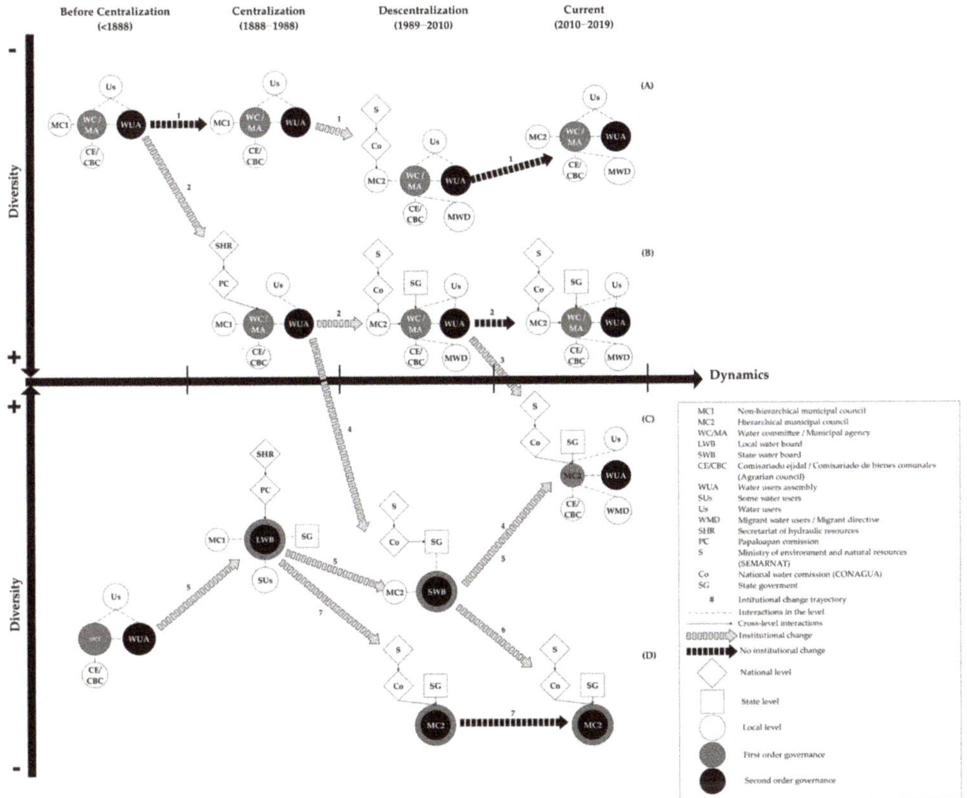

Figure 6. Intertwining the diversity, nestedness, and dynamics of stakeholders and institutions to understand the complexity of the governance of the domestic water supply system in Oaxaca, Mexico, under the Water Governance Complexity Framework. The current governance modes that resulted from the seven trajectories of institutional change are: (**A**) Non-nested community-based mode, (**B**) Nested community-based mode, (**C**) Nested hybrid mode, and (**D**) Nested municipal or hierarchical–bureaucratic mode. Source: prepared by the authors.

We found a diversity of stakeholders and evaluated how stakeholders and institutions conduct operations and decision making for the water supply system for domestic use at the local level. This diversity has allowed different arrangements of stakeholders in the first and second orders of governance. Likewise, these arrangements are differentiated by being nested or non-nested, given the types of cross-level interactions (which implies a hierarchical relationship between the jurisdictional levels) or interactions within the same level (in the community) between stakeholders. In this sense, we established four governance arrangements or modes that are presented in the water system for domestic use in Oaxaca, Mexico, at the local level (Figure 6):

- Non-nested community-based mode (Figure 6A). This mode is characterized by little or no cross-level interaction. Operations and decision making are conducted only between community stakeholders based on water committees and indigenous institutions (municipal agent and water user assembly). This mode was found in the communities of Santa Cruz Corunda, San Miguel Aztatla, El Enebro, and San Antonio Abad.
- Nested community-based mode (Figure 6B). This mode is similar to the non-nested community-based mode but with cross-level interactions mainly in the first order of governance with regard to financing hydraulic works with municipal, national, and state governments. This mode of governance was presented by Santiago Quiotepec, San Francisco Teopan, La Mexicana, and Santa Cruz Capulalpam.
- Nested hybrid mode (Figure 6C). This mode combines decision making between the stakeholders and institutions of the communities with municipal management based on national and state institutions (LAN). The nestedness occurs due to cross-level interactions of the municipality in the first order of governance regarding the financing, repair, and maintenance of the hydraulic infrastructure. This mode of governance was presented by all the municipal seats included in this study (Concepción Buenavista, Santiago Ihuitlan Plumas, Santa Magdalena Jicotlan, Santiago Tepetlapa, and San Juan de los Cues)
- Nested municipal or hierarchical–bureaucratic mode (Figure 6D). The governance of the water supply system for domestic use is conducted following the guidelines established by the LAN and Article 115. There is no participation in decision making on behalf of community stakeholders or water users through the assemblies. All management and decision making is conducted by the municipal operating body or the municipal council. This last mode of governance was not found in the communities in this study but is established according to the national institutions. This governance mode is the one that could become dominant in most municipalities, both in Oaxaca and in the rest of Mexico.

According to the number of stakeholders that make up these governance modes, the least diverse is the nested municipal mode, followed by the non-nested community-based mode. The two most diverse modes are the nested community-based and hybrid modes because they can incorporate all stakeholders of different jurisdictional levels into operations (Table 6, Figure 6). The substantial difference between the nested hybrid mode and the nested community-based mode can be seen in the second order of governance. In the nested hybrid mode, the municipality is involved in the first and second orders of governance. In the nested community-based mode, the municipality is only involved in the operations as a financier for hydraulic works that the community has decided it needs (Table 6). Secondly, according to national and state institutions, the hybrid nested mode emerges from the mix between indigenous institutions (e.g., water user assembly) and municipality management. National or state institutions do not consider this organizational operation of the water supply system.

Table 6. Diversity of stakeholders and institutions involved in the first and second orders of governance of the four modes of governance identified at the local level in Oaxaca and possibly in much of Mexico. Bold text indicates the entities responsible for operating and making decisions related to the water supply system for domestic use at the local level. Italic text indicates the stakeholders and institutions belonging to the local jurisdictional level, while Roman text indicates entities belonging to state and national jurisdictional levels. MC = municipal council, WC/MA = water committee/municipal agent, WUA = water users assembly, US = water users, CE/CBC = agrarian council, MWD = Migrant Water Users Directive, SG = State government, CO = National Water Commission, and S = Ministry of Environment and Natural Resources.

<table>
<tr><th rowspan="2">Modes of Governance</th><th colspan="2">Orders of Governance</th><th rowspan="2">Total of Different Stakeholders and Institutions Involved in Water Governance</th></tr>
<tr><th>First Order (Operativity)</th><th>Second Order (Institutional Arrangement)</th></tr>
<tr><td rowspan="2">Nested municipal mode</td><td colspan="2">MC</td><td rowspan="2">4</td></tr>
<tr><td>SG, S, and CO</td><td></td></tr>
<tr><td rowspan="3">Non-nested community-based mode</td><td>WC/MA</td><td>WC/MA and WUA</td><td rowspan="3">6</td></tr>
<tr><td>CE/CBC, MWD, and MC</td><td></td></tr>
<tr><td colspan="2">US</td></tr>
<tr><td rowspan="3">Nested community-based mode</td><td>WC/MA</td><td>WC/MA and WUA</td><td rowspan="3">9</td></tr>
<tr><td>SG, CO, S, CE/CBC, MWD, and MC</td><td></td></tr>
<tr><td colspan="2">US</td></tr>
<tr><td rowspan="3">Nested hybrid mode</td><td>MC</td><td>MC and WUA</td><td rowspan="3">9</td></tr>
<tr><td>SG, CO, S, CE/CBC, MWD, and MC</td><td></td></tr>
<tr><td colspan="2">US</td></tr>
</table>

The current diversity of stakeholders belonging to different jurisdictional levels (national, state, and local) and governance modes results from the trajectories of institutional changes related to water and the prevalence of indigenous institutions at the local level. The different trajectories of institutional change at the local level allow us to explain how the different modes of governance found were formed at this level. We identified seven trajectories of institutional change framed in four post-revolutionary periods, of which five were found in the 13 communities (numbers 1 to 7 represent specific trajectories of institutional change in Figure 6). All the trajectories of institutional change began in the period before centralization, starting with stakeholder and institutional arrangements based on the community through a water committee, municipal agent, water user assembly, or municipal council:

1. The water committee/municipal agent remained unchanged until the centralization period and subsequently became temporarily nested during the decentralization period. In the decentralization period, national, state, and municipal institutions constructed hydraulic infrastructure to bring water to homes. Additionally, migrant organization played an essential role in the financing, maintenance, and repair of hydraulic infrastructure. In the current period, a return to a governance structure like that present in the period before centralization is observed.
2. The water committee/municipal agent transitioned to a nested mode in the centralization period, where cross-level interactions with national government institutions were established. In the period of decentralization, nestedness was maintained although the national governmental institutions changed their names, structures, and functions, and state institutions were incorporated. Additionally, migrant organizations played an essential role in financing, maintaining, and repairing hydraulic infrastructure. In the current period, this has not changed.
3. The third trajectory of institutional change is similar to trajectory two. However, in the period from decentralization to the present, a change in stakeholders from the water committee/municipal agent to the municipal council in the first and second orders of governance was observed.

4. In this trajectory of institutional change, a change in stakeholders during the decentralization period occurred. The water committee/municipal agent disappeared, and the State Water Board (SWB) appeared in the arena of water governance at the local level. Subsequently, in the current period, the SWB disintegrated, and the municipal council takes its place in the first order of governance while water user assemblies retake the second order of governance (institutional arrangement).
5. This trajectory of institutional change is similar to trajectory 4. However, its beginnings prior to centralization are not due to a water committee/municipal agent but to a non-hierarchical municipal council. Another difference is that the governance structure changed to a Local Water Board (LWB) during the centralization period, and the municipal council lost power. In the decentralization period, the LWB was transformed into a SWB and national institutions went from being the main entities responsible to being advisors or financiers that provided technical support.
6. This trajectory of institutional change is similar to trajectory five up to the decentralization period. The difference with regard to trajectory five can be found in the current period. Instead of transitioning to a hybrid governance mode, a hierarchical–bureaucratic governance mode through the municipal council was adopted. The nested municipal council is the only one involved at the local level in the first and second orders of governance of the water supply system for domestic use.
7. This trajectory of institutional change is similar to trajectory six. However, during the period of decentralization, a swift change to the hierarchical–bureaucratic governance mode through the municipal council was observed instead of a transition to an SWB. The hierarchical–bureaucratic governance mode remains unchanged in the current period.

The last two trajectories (6 and 7) were built under the assumption that institutional changes at the national level thoroughly permeate the local level. The multiple trajectories of institutional change resulted from nestedness and the interplay of stakeholders belonging to the different jurisdictional levels. Nestedness plays an essential role in permeating institutional changes at higher jurisdictional levels or institutions to the local level. We can observe the contrasting effect in the non-nested communities of Santa Cruz Corunda, San Miguel Aztatla, El Enebro, and San Antonio Abad that have kept their local water institutions unchanged to date.

However, the interplay of stakeholders at the local level with governmental stakeholders belonging to higher jurisdictional levels in the different periods of change reflects non-linear interactions due to the resistance or adaptation of stakeholders or institutions at the local level. The interplay of stakeholders belonging to different jurisdictional levels helps to explain why institutional change at the national level did not permeate in the same way in all communities, despite communities being nested. For example, during centralization, the water supply systems for domestic use governed by water committees, indigenous institutions, or municipal councils would have disappeared, and only the LWBs would have prevailed. However, only two of the thirteen communities studied reported this change, reflecting the inability of the national government to take power away from stakeholders and local institutions over water matters. On the other hand, the resistance that local water institutions presented to change was imposed from the top down.

In the period of decentralization, there appears to be a return of water management power to the local water institutions present in the communities prior to centralization. Nevertheless, the return of power to local water institutions was accompanied by the establishment of a hierarchy and a homogenization of the operation of the water supply system for domestic use at the local level for the municipalities, concentrating operations, and decision making in a single stakeholder. For this reason, in the case of municipalities, we differentiate between municipal council 1 as non-hierarchical and municipal council 2 as hierarchical in Figure 6. Secondly, for the non-municipal water institutions (water committee, municipal agent, or water user assembly), a multilevel linkage with the municipality and state and national institutions was established. Finally, the national government maintains the right to alienation, while the right of exclusion is limited to the local water institutions, although mainly in the case of municipalities. During decentralization, differences among the trajectories of institutional change were due to the recovery of traditional institutions that existed before centralization in communities and the adaptation of traditional institutions to a new nested structuring of state and national organizations.

In the current period, recent institutional change, as seen with the Santiago Tepetlapa and San Juan de los Cues communities, responds to the endogenous drivers. Additionally, this institutional change may reflect the capacity of stakeholders to choose between the different institutional arrangements based on the community or municipality, or to generate new institutions that mix mechanisms coming from the municipality and indigenous institutions, such as in hybrid modes.

4. Discussion

Our results show that (1) legal pluralism is present due to the evolution and convergence of multiple formal (e.g., water, agrarian, and municipal laws, indigenous peoples rights, and environmental law) and informal (e.g., indigenous institutions such as those related to community use and customs, such as a municipal agent or water user assembly) institutions that co-exist at the local level. Even in the same legal system that regulates property rights over those of water, overlaps between national and state institutions at the local level are present along with concerns regarding the recognition of municipalities as responsible for the water supply system. (2) Legal pluralism has generated a great diversity of formal and informal stakeholders that are structured across multiple levels and are involved in various ways in the first (operative) and second orders of governance (institutional arrangements) of the water supply system for domestic use. (3) Diversity is associated with the four different modes of governance that exist and operate at the local level. The governance modes are determined by interactions (cross-level or within the jurisdictional level) between stakeholders in the first and second governance orders. They include the non-nested community-based mode, nested community-based mode, nested hybrid mode, and hierarchical–bureaucratic mode (municipality). (4) The municipality is nested by institutional design, unlike all indigenous communities and their respective modes (e.g., non-nested community-based mode) that manage the water system for domestic use through a municipal agent or water committee, and water user assembly. (5) The diversity and creation of governance modes in the water supply system for domestic use in Oaxaca and possibly in the rest of Mexico result from the seven different trajectories of institutional change. (6) The seven trajectories of institutional change result from nestedness and the interplay between local water institutions (e.g., municipalities and indigenous communities) and national and state government institutions during the centralization, decentralization, and current periods. Most of the diversity of stakeholders and institutions at the local level, modes of water governance, and trajectories of institutional change do not correspond to a single centralized plan but to the interplay of different stakeholders and institutions over time to secure water for households in the Mixtecan Alta region in Oaxaca, Mexico, and probably in the rest of the country.

These findings are consistent with empirical research, and in the case of nestedness and institutional change, they contribute new elements to our understanding of governance and institutional evolution [11]. The legal pluralism in water management in this study is consistent with what has been reported in previous studies regarding the growing legal pluralism in many countries due to the decentralization of water management [17–19], the existence of a multi-level process regarding the human right to water and sanitation that operates from the global to the local levels [20], the link between land and water rights in rural communities [18], and the recent recognition of indigenous rights and their traditional institutions [66]. Although it has not been viewed from a legal plural perspective, the diversity of stakeholders and governance modes agrees with what has been reported regarding the increasing number of stakeholders and novel institutional arrangements in the arena of water governance [67–69]. A novel institutional arrangement in the form of a hybrid mode has been suggested as being more likely to be present than other modes, such as the hierarchical–bureaucratic, market, or network modes [70]. The other modes of governance identified as non-nested community-based and nested community-based modes can be encased in the network mode [71]. However, we highlighted the critical role of the community and nestedness that reflects how the mode is affected by the system in which it is immersed [33]. Identifying nested and non-nested modes of governance allows us to identify new elements that shed light on how multi-level governance works. For example, the multi-level governance structure does not always imply a nesting of smaller organizations within larger organizations [72]. In the same structure, two subsystems can exist that function differently: one nested and another non-nested. From a functionalistic perspective, nestedness helps identify fragmented or coordinated governance in the water supply system [73]. We can observe that the existence of a non-nested mode of governance in the water supply system for domestic use in Oaxaca, Mexico, reflects the inability of national and state governmental institutions to reach specific communities, which implies a lack of multilevel coordination and a fragmented structure. It also gives us insight into the ability of a community to self-organize to meet the water demands of its inhabitants and to maintain both institutions and the household water supply over time. For its part, the trajectories of institutional change and the lack of change identified in this study contribute to filling the information gap regarding the longitudinal processes by which institutions are created and evolve [11], which were essential for explaining the structures of the modes of governance, particularly the emergence of hybrid modes.

From an overall perspective of the properties, legal pluralism, the diverse stakeholders and institutions, and nested or non-nested subsystems working in multilevel and dynamic properties due to institutional change establish the complexity of the governance of the water supply system for domestic use in Oaxaca. This complexity of the governance of the water supply system is confirmed if we consider the non-linear interactions (which are indirectly observed in the seven trajectories of institutional change) and emergence properties (new institutional arrangements such as those of hybrid modes) that were revealed due to the interrelatedness and complementarity between the diversity, nestedness, and dynamics properties. We empirically demonstrate that water governance becomes complex in structure and operation due to institutional evolution, with certain institutions aggregating, changing, adapting, and persisting over time while acting and interacting with stakeholders to supply water to people in different jurisdictional levels.

This study provides a replicable method to use the Water Governance Complexity Framework to understand the complexity of water systems. The framework considers diversity, nestedness, and dynamics at different scales (jurisdictional and temporal) and levels (national, state, and local) as well as periods of institutional change to address the complexity of water governance and the governance of the water supply system for domestic use in particular. This framework differs from the Kooiman IGF [14,39] by addressing complexity not only through a proxy (either legal pluralism [13] or nestedness [15], but as a property that encompasses diversity, nestedness, and dynamics. By themselves, each property provides an incomplete picture of water governance; however, taken together, they provide a more holistic understanding of the current structure and function of water governance, which can be complex. In the case of the governance of the domestic water supply system in Oaxaca, Mexico, we showed how this framework addresses the limitations of using a single variable or a set of separately viewed properties to understand water governance. Additionally, in contrast to the Kooiman framework that uses a more descriptive approach, we propose a variable-oriented approach that provides systematicity and replicability to describe the complexity of water governance in other regions. The Water Governance Complexity Framework joins recent efforts to advance our understanding of the past, present, and future of the institutions, and their interactions, and those of different frameworks, such as Power Polycentric Governance (PPG) [74] or a combination of PPG with other frameworks such as Institutional Analysis and Development (IAD) and the Socio-Ecological System (SES) Framework [11].

5. Conclusions

The purpose of this study was to approach the understanding of the complexity of water governance through three properties, namely diversity, nestedness, and dynamics, in jurisdictional and temporal scales and at different levels. For this, we built the Water Governance Complexity Framework based on some elements of the Kooiman Interactive Governance Framework [14,28]. We used the domestic water supply system in Oaxaca to show the operability of the proposed framework in addressing the complexity of water governance. We discovered that the properties of diversity, nestedness, and dynamics and their respective proxies when intertwined can provide good approximations of the non-linear interactions, emergence, and constant change that classify water governance as complex. The importance of this study and the Water Governance Complexity Framework is that it offers a way to understand the complexity of water governance due to historical processes without automatically assuming that water governance is complex, which limits and biases any conclusions or future improvement efforts.

The Water Governance Complexity Framework proposed in this study is not fixed. We recommend that the framework be used as a methodological guide by which new proxies can be incorporated to address the diversity, nestedness, and dynamics properties (e.g., the flow of knowledge and power dynamics) and new properties. New proxies may emerge from testing the framework in other regions or contexts and from advances in complexity theory. Additionally, a more refined level of analysis can be included in the framework, such as that at the individual level, which allows for a more profound understanding of complexity regarding how the stakeholders make decisions, how they implement specific actions, and how they interact with other stakeholders. Exploring the individual level will allow us to determine if the stakeholders of different jurisdictional levels participate in cross-level interactions from a legitimate non-hierarchical condition (different from how we assume nestedness) and the importance of leadership in inducing, resisting, or adapting to institutional change [75]. Despite the criticism of the theory of complexity when applied to social systems [76], we believe in its usefulness to diagnose the complexity of the water governance of any system, be it the water supply systems for domestic, agricultural, or industrial use. This practicality of the framework can help decision makers and practitioners generate a deeper understanding of water governance that allows for legal reforms, new laws, and new public policies to be created, along with changes to desirable models (e.g., polycentrism) based on the knowledge of current stakeholders and institutions and the historical context. This research reinforces the idea that water governance is complex, while inviting us to question this complexity and the elements and properties responsible for it.

Supplementary Materials: The following are available online at https://www.mdpi.com/2073-4441/13/20/2870/s1. Table S1. Structure of the survey used in this study. We employed four questions and their possible answers. In the answers, the "others" option was left not to limit the eventual appearance of different stakeholders at the local level. Table S2. Matrix used to note the answers obtained from interviewing stakeholders regarding cross-level and internal interactions in the communities to carry out ten activities of the first and second governance orders. Table S3. Matrix of binomial presence (1)/absence (0) data resulting from the collapse of the results obtained from all the interviewees and the ten activities of each community. Table S4. Results of the binomial presence/absence matrix of the communities that are municipal seats responsible for the water supply system for domestic use. This matrix was used to carry out the nesting analysis with the NODF (Nestedness based on the Overlap and Decreasing Fill) metric in NeD software. Table S5. Results of the binomial presence/absence matrix of the communities that are municipal agencies or that have a water committee responsible for the water supply system for domestic use. This matrix was used to carry out the nesting analysis with the NODF (Nestedness based on the Overlap and Decreasing Fill) metric in NeD software. Table S6. Open-ended questions of the semi-structured interviews were applied to elders and experts to identify possible institutional changes related to water for domestic use in rural communities.

Author Contributions: Conceptualization, F.G.-G.; methodology, F.G.-G., A.S.-A., G.H.-A., C.M.-A., M.A.M.-J. and D.M.-R.; formal analysis, F.G.-G.; investigation, F.G.-G. and G.H.-A.; resources, F.G.-G. and G.H.-A.; data curation, F.G.-G.; writing—original draft preparation, F.G.-G.; writing—review and editing, A.S.-A., G.H.-A., C.M.-A., M.A.M.-J. and D.M.-R.; supervision, A.S.-A.; project administration, F.G.-G.; funding acquisition, F.G.-G., A.S.-A., G.H.-A., C.M.-A., M.A.M.-J. and D.M.-R. All authors have read and agreed to the published version of the manuscript.

Funding: This research was funded by National Geographic Society (grant number EC-62227R-19) and The National Council of Science and Technology of Mexico (CONACYT) through Doctorate Grants awarded to the first author.

Institutional Review Board Statement: Not applicable.

Informed Consent Statement: Informed consent was obtained from all subjects involved in the study.

Data Availability Statement: The data presented in this study are available on request from the corresponding author.

Acknowledgments: We thank the communities of Oaxaca who participated in this study for their support and openness. We also thank the team of the Marine and Coastal Ecosystems Governance and Management Laboratory of the CIIDIR- Oaxaca unit of the Instituto Politécnico Nacional (IPN) for helping with the community surveys and for facilitating transportation. We are also deeply grateful for the suggestions made by three anonymous reviewers that helped us improve the scope of this work.

Conflicts of Interest: The authors declare no conflict of interest. The funders had no role in the design of the study; in the collection, analyses, or interpretation of data; in the writing of the manuscript; or in the decision to publish the results.

References

1. Kirschke, S.; Borchardt, D.; Newig, J. Mapping complexity in environmental governance: A comparative analysis of 37 priori-ty issues in German water management. *Environ. Policy Gov.* **2017**, *27*, 534–559. [CrossRef]
2. Kirschke, S.; Newig, J. Addressing complexity in environmental management and governance. *Sustainability* **2017**, *9*, 983. [CrossRef]
3. Pahl-Wostl, C. The implications of complexity for integrated resources management. *Environ. Model. Softw.* **2007**, *22*, 561–569. [CrossRef]
4. Pahl-Wostl, C.; Kranz, N. Water governance in times of change. *Environ. Sci. Policy* **2010**, *13*, 567–570. [CrossRef]
5. Cox, M. Applying a Social-Ecological System Framework to the Study of the Taos Valley Irrigation System. *Hum. Ecol.* **2014**, *42*, 311–324. [CrossRef]
6. McGinnis, M.D.; Ostrom, E. Social-ecological system framework: Initial changes and continuing challenges. *Ecol. Soc.* **2014**, *19*. [CrossRef]
7. Ostrom, E. A general framework for analyzing sustainability of social-ecological systems. *Science* **2009**, *325*, 419–422. [CrossRef]
8. Chaffin, B.C.; Gunderson, L.H. Emergence, institutionalization and renewal: Rhythms of adaptive governance in complex social-ecological systems. *J. Environ. Manag.* **2016**, *165*, 81–87. [CrossRef]
9. Schröder, N.J.S. The lens of polycentricity: Identifying polycentric governance systems illustrated through examples from the field of water governance. *Environ. Policy Gov.* **2018**, *28*, 236–251. [CrossRef]
10. Carlisle, K.M.; Gruby, R.L. Why the path to polycentricity matters: Evidence from fisheries governance in Palau. *Environ. Policy Gov.* **2018**, *28*, 223–235. [CrossRef]
11. Epstein, G.; Morrison, T.H.; Lien, A.; Gurney, G.G.; Cole, D.H.; Delaroche, M.; Villamayor-Tomas, S.; Ban, N.; Cox, M. Advances in understanding the evolution of institutions in complex social-ecological systems. *Curr. Opin. Environ. Sustain.* **2020**, *44*, 58–66. [CrossRef]
12. Pahl-Wostl, C.; Holtz, G.; Kastens, B.; Knieper, C. Analyzing complex water governance regimes: The Management and Transition Framework. *Environ. Sci. Policy* **2010**, *13*, 571–581. [CrossRef]
13. Jentoft, S. Limits of governability: Institutional implications for fisheries and coastal governance. *Mar. Policy* **2007**, *31*, 360–370. [CrossRef]
14. Kooiman, J. *Governing as Governance*; SAGE: London, UK, 2003.
15. Bavinck, M.; Kooiman, J. Applying the Governability Concept in Fisheries - Explorations from South Asia. In *Governability of Fisheries and Aquaculture: Theory and Applications*; Bavinck, M., Chuenpagdee, R., Jentoft, S., Kooiman, J., Eds.; Springer: Dordrecht, The Netherlands, 2013; Volume 7, pp. 131–153.
16. Manson, S.M. Simplifying complexity: A review of complexity theory. *Geoforum* **2001**, *32*, 405–414. [CrossRef]
17. Tamanaha, B.Z. Understanding Legal Pluralism: Past to Present, Local to global. *Syd. Law Rev.* **2007**, *30*, 375–411. [CrossRef]

18. Meinzen-Dick, R.; Nkonya, L. Understanding legal pluralism in water rights: Lessons from Africa and Asia. In *Community-Based Water Law and Water Resource Management Reform in Developing Countries*; van Koppen, B., Giordano, M., Butterworth, J., Eds.; CAB International: Wallingford, UK, 2007.
19. McCord, P.; Dell'Angelo, J.; Baldwin, E.; Evans, T. Polycentric transformation in Kenyan water governance: A dynamic analysis of institutional and social-ecological change. *Policy Stud. J.* **2017**, *45*, 633–658. [CrossRef]
20. Obani, P.; Gupta, J. Legal pluralism in the area of human rights: Water and sanitation. *Curr. Opin. Environ. Sustain.* **2014**, *11*, 63–70. [CrossRef]
21. Kooiman, J. Exploring the Concept of Governability. *J. Comp. Policy Anal. Res. Pract.* **2008**, *10*, 171–190. [CrossRef]
22. Kooiman, J.; Bavinck, M. Theorizing Governability—The Interactive Governance Perspective. In *Governability of Fisheries and Aquaculture: Theory and Applications*; Bavinck, M., Chuenpagdee, R., Jentoft, S., Kooiman, J., Eds.; Springer: Dordrecht, The Netherlands, 2013; Volume 7, pp. 9–31.
23. Tortajada, C. Water governance: Some critical issues. *Int. J. Water Resour. Dev.* **2010**, *26*, 297–307. [CrossRef]
24. Lewis, K. *Water Governance for Poverty Reduction: Key Issues and the UNDP Response to Millennium Development Goals*; United Nations Development Programme: New York, NY, USA, 2004.
25. North, D.C. Institutions. *J. Econ. Perspect.* **1991**, *5*, 97–112. [CrossRef]
26. Page, S. On diversity and complexity. In *Diversity and Complexity*; Page, S., Ed.; Princeton University Press: Princeton, NJ, USA, 2011; pp. 16–53.
27. Ostrom, E. Understanding the diversity of structured human interactions. In *Understanding Institutional Diversity*; Ostrom, E., Ed.; Princeton University Press: Princeton, NJ, USA; Oxford, UK, 2005; pp. 3–29.
28. Kooiman, J.; Bavinck, M.; Chuenpagdee, R.; Mahon, R.; Pullin, R. Interactive governance and governability—An introduction. *J. Transdiscipl. Environ. Stud.* **2008**, *7*, 1–11.
29. Kingston, C.; Caballero, G. Comparing theories of institutional change. *J. Inst. Econ.* **2009**, *5*, 151. [CrossRef]
30. Gibson, C.C.; Ostrom, E.; Ahn, T.K. The concept of scale and the human dimensions of global change: A survey. *Ecol. Econ.* **2000**, *32*, 217–239. [CrossRef]
31. Bavinck, M.; Chuenpagdee, R.; Jentoft, S.; Kooiman, J. Social Justice in the Context of Fisheries—A Governability Challenge. In *Governability of Fisheries and Aquaculture: Theory and Applications*; Chuenpagdee, R., Jentoft, S., Kooiman, J., Eds.; Springer: Dordrecht, The Netherlands, 2013; Volume 7, pp. 45–66.
32. Lautze, J.; De Silva, S.; Giordano, M.; Sanford, L.; De Silva, S. Putting the cart before the horse: Water governance and IWRM. *Nat. Resour. Forum.* **2011**, *35*, 1–8. [CrossRef]
33. Blavoukos, S.; Bourantonis, D. Nested Institutions. In *Palgrave Handbook of Inter-Organizational Relations in World Politics*; Koops, J.A., Biermann, R., Eds.; Palgrave Macmillan: London, UK, 2017; pp. 303–317.
34. Mariani, M.S.; Ren, Z.M.; Bascompte, J.; Tessone, C.J. Nestedness in complex networks: Observation, emergence, and implications. *Phys. Rep.* **2019**, *813*, 1–90. [CrossRef]
35. Levin, S.A. The problem of pattern and scale in ecology. *Ecology* **1992**, *73*, 1943–1967. [CrossRef]
36. Ostrom, E. *Governing the Commons: The Evolution of Institutions for Collective Action (Political Economy of Institutions and Decisions)*; Cambridge University Press: Cambridge, UK, 1990.
37. Theodori, G.L. Community and community development in resource-based areas: Operational definitions rooted in an interactional perspective. *Soc. Nat. Resour.* **2005**, *18*, 661–669. [CrossRef]
38. Fox, J.A.; Aranda, J. *Decentralization and Rural Development in Mexico Community Participation in Oaxaca's Municipal Funds Programs*; Center for U.S.-Mexican Studies, University of California: San Diego, CA, USA, 1996.
39. Kooiman, J.; Chuenpagdee, R. Governance and Governability. In *Fish for Life: Interactive Governance for Fisheries*; Kooiman, J., Bavinck, M., Jentoft, S., Pullin, R., Eds.; Amsterdam University Press: Amsterdam, The Netherlands, 2005; pp. 325–350.
40. Naderifar, M.; Goli, H.G.; Fereshteh, G. Snowball Sampling: A Purposeful Method of Sampling in Qualitative Research. *Strides Dev. Med. Educ.* **2017**, *14*, e67670. [CrossRef]
41. Almeida-Neto, M.P.; Guimarães, R.J.; Loyota, R.D.; Ulrich, W. A consistent metric for nestedness analysis in ecological systems: Reconciling concept and measurement. *Oikos* **2008**, *117*, 1227–1239. [CrossRef]
42. Strona, G.; Galli, P.; Seveso, D.; Montano, S.; Fattorini, S. Nestedness for Dummies (NeD): A User-Friendly Web Interface for Exploratory Nestedness Analysis. *J. Stat. Softw.* **2014**, *59*, 1017–1025. [CrossRef]
43. Ulrich, W.; Gotelli, N.J. Null model analysis of species nestedness patterns. *Ecology* **2007**, *88*, 1824–1831. [CrossRef]
44. Galindo-Escamilla, E.; Palerm-Viqueira, J. Pequeños sistemas de agua potable: Entre la autogestión y el manejo municipal en el estado de Hidalgo, México. *Agric. Soc. Y Desarro.* **2007**, *4*, 127–145.
45. Pineda, P.N. La política urbana de agua potable en México: Del centralismo y los subsidios a la municipalización, la autosuficiencia y la privatización. *Reg. Soc.* **2002**, *14*, 41–70. [CrossRef]
46. Sandré, O.I.; Sánchez, M. *El eslabón Perdido. Acuerdos, Convenios, Reglamentos y Leyes Locales de Agua en México (1593–1935)*; Centro de Investigaciones y Estudios Superirores en Antropologia Social, Publicaciones de la Casa Chata: D. F., México, Mexico, 2011; pp. 11–54.
47. Aboites, A.L. *El agua de la Nación. Una Historia Política de México (1888–1946)*; Centro de Investigaciones y Estudios Superiores en Antropología Social: D. F., México, Mexico, 1998.
48. Rolland, L.; Vega, Y. La gestión del agua en México. *Polis* **2010**, *6*, 155–188.

49. Collado, M.J. Entorno de la Provisión de los Servicios Públicos de agua Potable en México. In *El Agua Potable en México. Historia Reciente, Actores, Procesos y Propuestas*; Olivares, R., Sandoval, R., Eds.; Asociación Nacional de Empresas de Agua y Saneamiento, A.C: D. F., México, Mexico, 2008; pp. 3–28.
50. Escobar, O.A. Manejo del Agua en México. Bosquejo de la evolución institucional federal 1926–2008. In *Semblanza Historica del Agua en México*; Comisión Nacional del Agua, Ed.; Secretaria de Medio Ambiente y Recursos Naturales: D. F., México, Mexico, 2009; pp. 61–73.
51. Mathews, A.S. Building the town in the country: Official understandings of fire, logging and biodiversity in Oaxaca, Mexico, 1926–2004. *Soc. Anthropol.* **2006**, *14*, 335–359. [CrossRef]
52. DOF. *Constitución Política de los Estados Unidos Mexicanos*; Diario Oficial de la Federación: D. F., México, Mexico, 1917; Volume 4, pp. 1–418.
53. DOF. *Ley de Aguas Nacionales*; Diario Oficial de la Federación: D. F., México, Mexico, 1992; pp. 1–112.
54. DOF. *Ley Agraria*; Diario Oficial de la Federación: D. F., México, Mexico, 1992; pp. 1–55.
55. DOF. *Ley General del Equilibrio Ecologico y Proteccion al Medio Ambiente*; Diario Oficial de la Federación: D. F., México, Mexico, 1988; pp. 1–268.
56. DOF. *Decreto por el que se Crea la Comisión Nacional del Agua Como Órgano Admisnistrativo Deseconcentrado de la Secretaría de Agricultura y Recursos Hidráulicos*; Diario Oficial de la Federación: D. F., México, Mexico, 1989; Volume 11, p. 2.
57. DOF. *Acuerdo Por el que se Determina la Circunscripción Territorial de los Organismos de Cuenca de la Comisión Nacional del Agua*; Diario Oficial de la Federación: D. F., México, Mexico, 2010. Available online: http://dof.gob.mx/nota_detalle.php?codigo=5137623&fecha=01/04/2010 (accessed on 10 February 2021).
58. DOF. *Decreto por el que se Declara Reformado y Adicionado el Artículo 115 de la Constitución Política de los Estados Unidos Mexicanos*; Diario Oficial de la Federación: D. F., México, Mexico, 1999. Available online: http://www.dof.gob.mx/nota_detalle.php?codigo=4958409&fecha=23/12/1999 (accessed on 4 March 2021).
59. POEO. *Ley de Agua Potable y Alcantarillado Para el Estado de Oaxaca*; Periodico Oficial del Estado de Oaxaca: Oaxaca, Mexico, 1993; Volume 22, pp. 373–388.
60. Gumeta-Gómez, F.; Durán, E.; Hinojosa-Arango, G. Gobernanza del agua para uso doméstico. *H_2O Gestión del Agua* **2019**, *26*, 22–26.
61. Birrichaga, D. Legislación entorno al agua, siglos XIX y XX. In *Semblanza Historica del Agua en México*; Comisión Nacional del Agua, Ed.; Secretaria de Medio Ambiente y Recursos Naturales: D. F., México, Mexico, 2009; pp. 43–59.
62. Zetina, R.M.D.C.; García, P.R.; Rangel, G.E. Administración del agua y los recursos de la nación: La Junta Federal de Mejoras Materiales, Ciudad Juárez, Chihuahua, 1931–1936. *Región Y Soc.* **2017**, *29*, 1931–1936. [CrossRef]
63. CONAGUA. *Manual de Agua Potable, Alcantarillado y Saneamiento. Integración de un Organismo Operador*; Secretaria del Medio ambiente y Recursos Naturales: D. F., México, Mexico, 2007; pp. 1–3.
64. Jimenez, B.; Marín, L. *El Agua en Mexico, Vista Desde la Academia*; Academia Mexicana de Ciencias: D. F., México, Mexico, 2004.
65. Wilder, M.; Romero, L.P. Paradoxes of decentralization: Water reform and social implications in Mexico. *World Dev.* **2006**, *34*, 1977–1995. [CrossRef]
66. Anaya, J. Indigenous law and its contribution to global pluralism. *Indig. LJ* **2007**, *6*, 3.
67. Gupta, J.; Pahl-Wostl, C.; Zondervan, R. 'Glocal' water governance: A multi-level challenge in the anthropocene. *Curr. Opin. Environ. Sustain.* **2013**, *5*, 573–580. [CrossRef]
68. Gawel, E.; Bernsen, K. Globalization of water: The case for global water governance. *Nat. Cult.* **2011**, *6*, 205–217. [CrossRef]
69. Pahl-Wostl, C. *Water Governance in the Face of Global Change. From Understanding to Transformation*; Springer: Osnabrük, Germany, 2015.
70. Pahl-Wostl, C. The role of governance modes and meta-governance in the transformation towards sustainable water governance. *Environ. Sci. Policy* **2019**, *91*, 6–16. [CrossRef]
71. Meuleman, L. *Public Management and the Metagobernance of Hierarchies, Networks and Markets*; Springer: Berlin/Heidelberg, Germany, 2008.
72. Sjah, T.; Baldwin, C. Options for future effective water management in lombok: A multi-level nested framework. *J. Hydrol.* **2014**, *519*, 2448–2455. [CrossRef]
73. Pahl-Wostl, C.; Knieper, C. The capacity of water governance to deal with the climate change adaptation challenge: Using fuzzy set Qualitative Comparative Analysis to distinguish between polycentric, fragmented and centralized regimes. *Glob. Environ. Chang.* **2014**, *29*, 139–154. [CrossRef]
74. Morrison, T.H. Evolving polycentric governance of the Great Barrier Reef. *Proc. Natl. Acad. Sci. USA* **2017**, *114*, E3013–E3021. [CrossRef]
75. Folke, C.; Hahn, T.; Olsson, P.; Norberg, J. Adaptive governance of social-ecological systems. *Annu. Rev. Environ. Resour.* **2005**, *30*, 441–473. [CrossRef]
76. Stewart, P. Complexity theories, social theory, and the question of social complexity. *Philos. Soc. Sci.* **2001**, *31*, 323–360. [CrossRef]

MDPI

Article

Spatiotemporal Changes in Mulberry-Dyke-Fish Ponds in the Guangdong-Hong Kong-Macao Greater Bay Area over the Past 40 Years

Wenxin Zhang [1], Zihao Cheng [1], Junliang Qiu [2], Edward Park [3,4,5], Lishan Ran [6], Xuetong Xie [1] and Xiankun Yang [1,7,*]

1 School of Geography and Remote Sensing, Guangzhou University, Guangzhou 510006, China; wenxinzhang@e.gzhu.edu.cn (W.Z.); 2980264379@e.gzhu.edu.cn (Z.C.); xiexuetong@gzhu.edu.cn (X.X.)
2 Department of Land, Environment, Agriculture and Forestry, University of Padova, Agripolis, viale dell'Università 16, 35020 Legnaro (PD), Italy; junliang.qiu@studenti.unipd.it
3 National Institute of Education, Nanyang Technological University, Singapore 637616, Singapore; edward.park@nie.edu.sg
4 Asian School of the Environment, Nanyang Technological University, Singapore 639798, Singapore
5 Earth Observatory of Singapore, Nanyang Technological University, Singapore 639798, Singapore
6 Department of Geography, The University of Hong Kong, Hong Kong, China; lsran@hku.hk
7 Rural Non-Point Source Pollution Comprehensive Management Technology Center of Guangdong Province, Guangzhou University, Guangzhou 510006, China
* Correspondence: yangxk@gzhu.edu.cn

Citation: Zhang, W.; Cheng, Z.; Qiu, J.; Park, E.; Ran, L.; Xie, X.; Yang, X. Spatiotemporal Changes in Mulberry-Dyke-Fish Ponds in the Guangdong-Hong Kong-Macao Greater Bay Area over the Past 40 Years. *Water* **2021**, *13*, 2953. https://doi.org/10.3390/w13212953

Academic Editor: Athanasios Loukas

Received: 16 August 2021
Accepted: 17 October 2021
Published: 20 October 2021

Publisher's Note: MDPI stays neutral with regard to jurisdictional claims in published maps and institutional affiliations.

Copyright: © 2021 by the authors. Licensee MDPI, Basel, Switzerland. This article is an open access article distributed under the terms and conditions of the Creative Commons Attribution (CC BY) license (https://creativecommons.org/licenses/by/4.0/).

Abstract: Mulberry-dyke-fish pond ecosystems are a representative traditional eco-agriculture in the Guangdong–Hong Kong–Macao Greater Bay Area (GBA). Investigations about the changes in the systems and their relevant water environments under the background of rapid urbanization can provide valuable information to formulate sustainable protection and development strategies. Using the Landsat images obtained after 1986, this study combined supervised classification and visual interpretation approaches, as well as water intensity index and synthesized index to identify the spatial patterns of changes in the ponds in the GBA over the past 40 years. The results indicated that during the period 1986–2013, the total surface area of the ponds in the GBA increased significantly and peaked in 2013 with a total increase of 84.63%; After that, the total surface area showed a downward trend with a total decrease of approximately 31.34%. The year of 2013 was identified as the milestone of the changes. The results proved that human activities have continuously influenced the spatial distribution and size of fish ponds in the past 40 years. The fish ponds had transformed from near-natural ponds with different sizes and a near-natural random distribution in the early stage into an artificial distribution and an artificial shape. Land use changes, industrial transfer, Government guidance and financial motives were the major drivers to the changes. If no effective measures are taken, this shrinking trend in the ponds will remain in the future.

Keywords: mulberry-dyke-fish pond ecosystem; spatial evolution analysis; remote sensing; Guangdong–Hong Kong–Macao Greater Bay Area

1. Introduction

The mulberry-dyke-fish pond ecosystem, developed by ancient Chinese farmers 2500 years ago with complex irrigation and drainage design, is an artificial eco-agriculture system which are mostly found in the Yangtze River Delta and Pearl River Delta (PRD). Conducive to the cultivation of mulberry-dyke trees, silk rearing, fish and poultry farming, the ecosystem plays an important role in energy circulation and the ecological environment protection [1]. Compared to other agricultural systems, this system has better economic and ecological performance, with advantages in regulation of droughts and floods, stable high outputs, and easy operation [2]. Currently, this traditional agriculture system is believed under a crisis of extinction caused by the outflow of rural population and the fast expansion of cities and towns, especially in the GBA, which now is one of the most prosperous regions in China. One of the manifestations is that, as part of the GBA, the Pearl River Delta (PRD) region is one of China's most important urban agglomerations with the fastest urbanization rate. Profited from the implementation of China's reform and opening-up policy since 1978, the area of built-up land in the PRD reached 434,570 hm^2 from 1988 to 1998 alone and enlarged by 1.5 times in a ten-year period. This process undoubtedly occupied ponds [1], making them more fragile because of the associated polluted water [3,4]. Moreover, views about the shrinking in the ponds could be summarized into two aspects. First are policy changes. The current average annual growth rate of national agricultural production needs to be maintained at 4.6% to meet the food needs of 22% of the global population, while the output by the ponds is not comparable with large-scale modern agriculture. The traditional pond ecosystem is thus being abandoned by farmers. Second is rural nonpoint and mini-point source pollution. Pollutants from widely used synthetic fertilizers cause various effects to the pond ecosystem, exacerbating material and energy flows within the ecosystem, thus weakening its ecosystem service. Compared with other large waterbodies, ponds are less capable of pollutant dilution, leading them to be abandoned. Under these circumstances, there have been increasing awareness to implement conservation policies for the pond ecosystem preservation. Measures such as numerical assessments, pond inventory mapping and pond ecosystem monitoring through IoT-based devices have been applied [5].

Previous studies on the ponds in the GBA mainly focused on its ecological functions, such as applications of energy theories to make synthetical and quantitative analyses based on energy structure and indices [1]. Their historical development, agricultural heritage and landscape patterns were also investigated. The pond ecosystem, as an important part of the agricultural heritage systems, has prominent agricultural heritage values for enriching production diversity and biological diversity in the GBA [6]. It is often recognized as a reflection of the harmonious coexistence of man and nature, demonstrating circular economy and ecological civilization ideas in China [7]. However, the current unclear status about the ponds has restricted these investigations.

For the monitoring of temporal and spatial changes in fish ponds, a combination of Landsat images from 2000, 2005, and 2015 has been employed to analyze them in the Foshan City in western GBA, accompanied by a similar study in Foshan, to analyze the spatial pattern changes in 1988, 1998, and 2006 [8,9]. Located in the central part of GBA, Zhongshan City has also been studied about the dynamics of its fish ponds using Landsat images in 1990, 2000 and 2013 [10]. Regarding the whole GBA, Keyhole images from 1964 and 1976 were combined with Landsat images from 1988, 2000 and 2012 to detect its ponds' spatial changes [10,11]. The dynamics in fish ponds have been almost released in this study, yet the time-series was still relatively short in comparison with such a long-term (more than 50 years) urbanization process in the GBA. As a result, previous studies did not fully reflect the spatiotemporal changes at a relatively complete scale. The long-term spatial dynamics found in the fish ponds is still unclear.

To respond to the research gap, based on the Landsat images obtained in the period 1986–2019, this study conducted the investigated the long-term spatiotemporal changes in fish ponds and their landscape dynamics to reveal fish ponds' historical development in the GBA. The study results will be valuable for a more comprehensive understanding of the water ecosystem dynamic development in the GBA in past decades. In addition, influencing factors were analyzed to provide an accurate reference for decision making on pond ecosystem restoration and conservation.

2. Data and Methods

2.1. Study Area

The Guangdong–Hong Kong–Macao Greater Bay Area (GBA) is located in south-central China, composed of the "9 + 2 urban agglomeration" which is composed of two special administrative regions of Hong Kong and Macau and 9 cities from central Guangdong Province (namely, Guangzhou, Shenzhen, Zhuhai, Foshan, Huizhou, Dongguan, Zhongshan, Jiangmen and Zhaoqing). It ranges between 24.4°~21.5° N, 111.4°~115.4° E, with a total area of about 56,000 km^2 (Figure 1). The population of GBA reached approximately 70 million in 2017, with USD 1.51 trillion GDP and urbanization rate of 85.20% excluding Hong Kong and Macau. In some of the cities such as Shenzhen and Foshan, the urbanization even exceeded 90% [12]. The GBA is currently one of the strongest economic vitality regions in China. Not only that, but it is also one of the most typical areas of fish pond adoption in China. However, with the fast industrialization and urbanization in the GBA, fish ponds have been seriously degraded in past decades [13–15]. In recent years, with the further acceleration of the transition from rural population to urban population, the built-up area in the GBA has grown at an annual rate of 3.35% from 2000 to 2015. All high-density populated areas (>10 inhabitants/900 m^2) are located in built-up areas, and the average population density in rural areas has also decreased at a rate of 1% per year [16]. Accompanied by this phenomenon, decreasing farmers in the GBA would be a potential threat for fish pond management. Thus, a comprehensive analysis of the long-term dynamics of fish ponds and the impacts of urbanization on their changes is indispensable in this context.

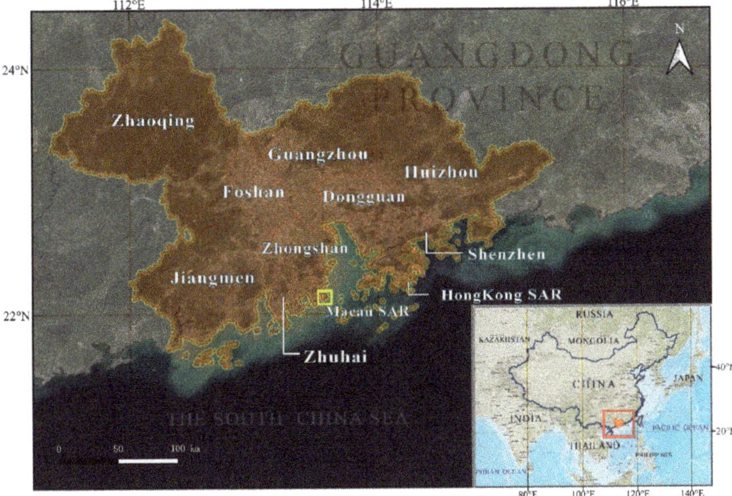

Figure 1. Administrative divisions of the Guangdong–Hong Kong–Macao Greater Bay Area (GBA). The study area covered the entire GBA. Because there are only a few fish ponds in Macau SAR, they were excluded in following investigations.

2.2. Data Collection and Image Analysis

2.2.1. Data Source

In total, 120 scenes of cloud-free Landsat images in the years1986, 1988, 1991, 1994, 1996, 1999, 2006, 2013, 2015 and 2019, were selected for fish pond delineation. Landsat-5 TM imagery were adopted from the 1980s to the 2000s, and Landsat-8 OLI imagery were used for the 2010s. Both of the two datasets have the characteristics of multiple bands and high spectral resolution, equipped with rich information and possessed 30 m spatial resolution [17]. The near-infrared band is extremely sensitive to vegetation and water bodies. Preprocessing processes include radiation correction, geometric correction, mosaic and extraction by GBA boundary were performed by the Landsat images preprocessing module provided by ENVI 5.3.

Because different regions have different degrees of urbanization. Dividing the entire study area of GBA into different sub-regions can enable a better understanding of the degrees of urbanization and their impact on pond variations in different regions. Therefore, to facilitate further analysis, the 9 prefectural cities and Hongkong Special Administrative Region (since the fish ponds in the Macau Special Administrative Region has disappeared for years, it was noted included in the study) were further divided into 25 research units (Figure 2). They included 5 units in the prefectural city of Foshan, namely, (1) Sanshui, (2) Nanhai, (3) Shunde, (4) Gaoming, and (5) Chancheng districts. Besides, 4 units are located in the city of Guangzhou, namely, (6) Panyu, (7) Baiyun and Huadu districts (combined as Unit A), (8) Nansha, (9) Yuexiu, Liwan, Haizhu, Tianhe, Huangpu, Zengcheng and Conghua districts (Combined as Unit B). The other 5 individual prefectural cities (i.e., Dongguan (10), Huizhou (11), Shenzhen (12), Zhuhai (13), Zhongshan (14)) and Hong Kong SAR (15) were set as another 6 units. Then, 4 units in the prefectural city of Zhaoqing: (16) Gaoyao district, (17) Sihui County, (18) Duanzhou district and Dinghu district (combined as Unit C), (19) Deqing County, Huaiji County, Fengkai County, Guangning County (combined as Unit D). Further, there are 6 units in the prefectural city of Jiangmen: (20) Heshan city, (21) Xinhui district, (22) Taishan city, (23) Jianghai district, (24) Pengjiang district, (25) Kaiping city and Enping city (combined as Unit E).

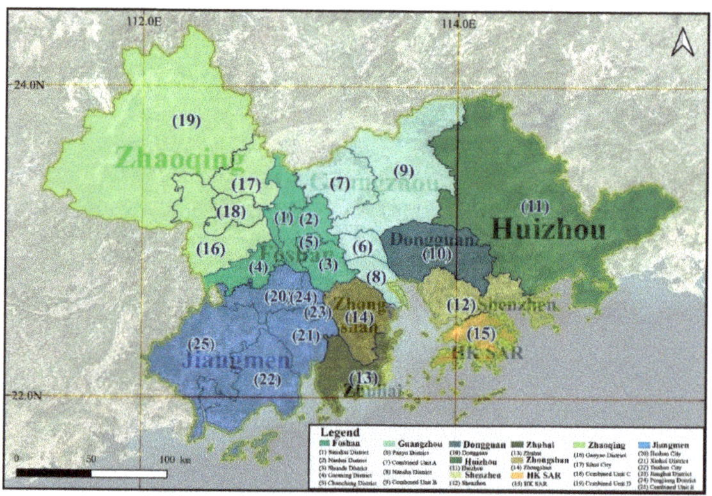

Figure 2. The location of 25 research units divided.

To ensure image classification accuracy, field investigations to collect fish ponds' spectral information using portable ASD FieldSpec were conducted in cities such as Foshan and Guangzhou where fish ponds are commonly located (Figure 3). The field-collected spectrum curves of fish ponds were added as part of the spectral library for classification. The accuracy assessment was proposed by building the confusion matrix comparing the extraction results and the accuracy assessment samples we designed based on Landsat images, combined with Google Earth historical images, field survey data and historical land use data, etc. Then, the Kappa coefficient calculated form confusion matrix was adopted to examine the accuracy of the classification results (Figure 4).

Figure 3. Photos of field investigations for the collection of ponds' spectral information. The photos were taken by the authors.

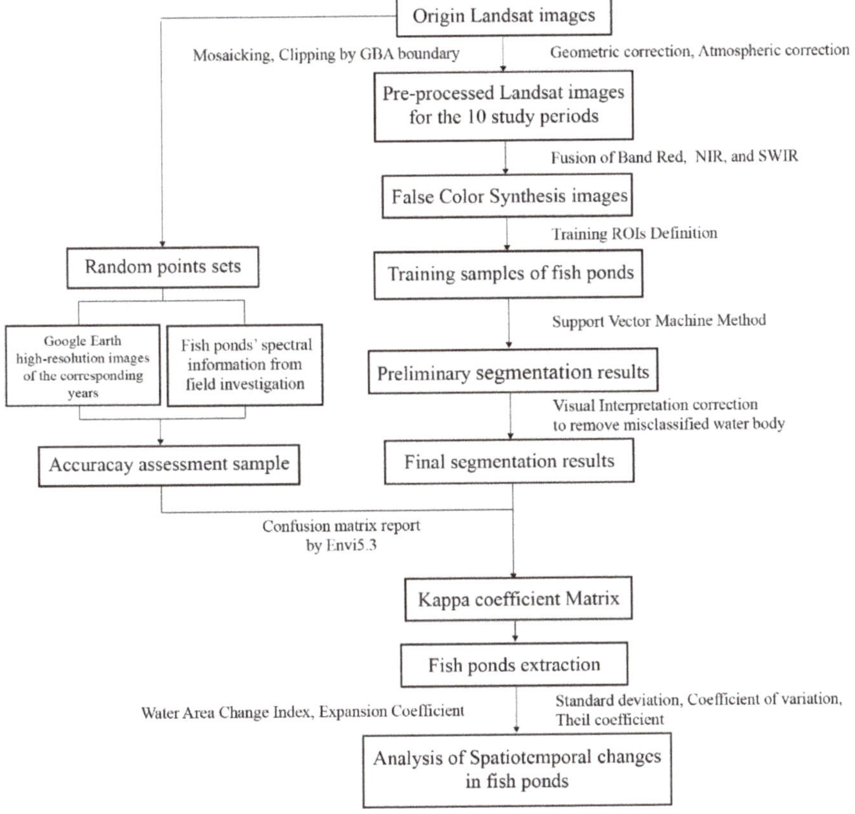

Figure 4. Flowchart for image classification and analysis.

2.2.2. Image Classification

After the pre-processing of the images, the standard images in the study periods were obtained (Figure 5) to perform classification by the support vector machine algorithm. Figure 6 showed the representative patterns of waterbodies in Landsat images. The areas in blue or black colors with a clear rectangular or square boundary are fish ponds, while the rivers that extends in strip shapes.

Figure 5. Pre-processed Landsat images for the 10 study periods.

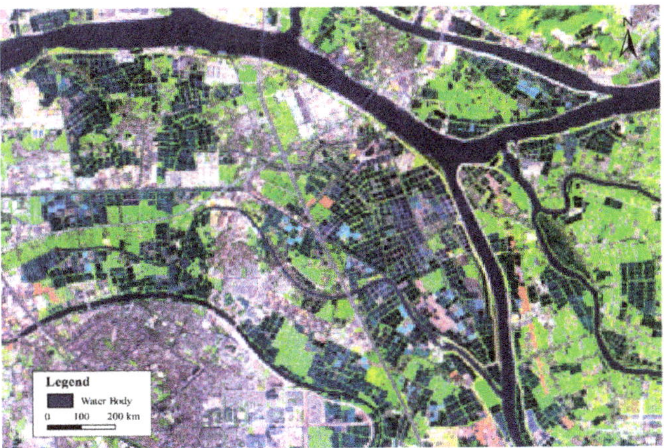

Figure 6. Representative morphological features of fish ponds in Landsat images.

Some initial information can be obtained from these pre-processed images (Figure 5). Regarding image quality, it can be seen that the image quality of the early Landsat satellite images is somewhat lower than the recently acquired Landsat OLI images. It can be clearly seen that the early satellite images were fuzzier than the latest ones, indicating more noises in these images. In addition, it was observed that the urbanization in the GBA has experienced rapid development. The built-up area (in purple color) has substantially expanded.

2.3. Methods

2.3.1. Quantitative Evaluation Method

The standard deviation (*SD*) is the square of the deviation between the standard value of the fish pond size and the average area in each period. It was used to reflect the dispersion of the ponds within each period. Therefore, a larger *SD* indicates a greater difference in pond size. The coefficient of variation (*CV*) is the ratio of the standard deviation (*SD*) of the pond area to the corresponding average in each period. A larger *V* suggests a higher relative difference in pond size within each study unit [18].

The Theil coefficient (*T*) indicates the regional inequality and is used to quantify the degree of difference in individual pond sizes. A larger *T* indicates a larger inequality in pond size. In the formula, x_i is the individual pond area; x_0 is the mean value of the pond area. *N* is the total number of ponds.

$$SD = \sqrt{\frac{1}{N}\sum_{i=1}^{N}(x_i - x_0)} \tag{1}$$

$$CV = \frac{SD}{x_0} \tag{2}$$

$$T = \frac{1}{N}\sum_{i=1}^{N} ln\frac{x_0}{x_i} \tag{3}$$

2.3.2. Water Area Change Index (*W*)

The water area change intensity index indicates the rate of the areal change of the ponds in a specific period.

$$W = 100\% \times \frac{S_b - S_a}{S_a \times \Delta t}, \tag{4}$$

where, S_a and S_b represent the area of the ponds in the previous period a and the current period b, respectively, Δt represents the timespan between periods a and b.

2.3.3. Expansion Coefficient (*E*)

Expansion Coefficient is an important quantitative indicator to measure the development and change in fish ponds:

$$E = \prod_{i=1}^{3} e_i, \tag{5}$$

where, *E* represents the expansion coefficient, which is obtained by multiplying the internal structure transition coefficient (e_1), the spatial structure transition coefficient (e_2) and the expansion coefficient (e_3).

Internal structure transition coefficient: $e_1 = |P_t - P_{t-1}|$, e_1 in the formula represents the relative areal change of ponds. P_t represents the areal percentage of ponds in the total area in current period, and P_{t-1} represents the one in the previous period.

Spatial structure transition coefficient: $e_2 = |S_t - S_{t-1}|$, S_t and S_{t-1} respectively represent the total fish pond area in the last and the first periods.

Expansion coefficient: $e_3 = \left|\sqrt[t]{S_t/S_0} - 1\right|$. e_3 is for the gradient change rate of fish ponds; S_t, and S_0 respectively represent total pond area for the last and first periods. *T* is the research period.

In order to discriminate the patterns of the pond changes of each study unit, the 26 study units were generally divided into 4 patterns: fast growing ($E > 40$), weak growing ($40 > E > 1$), stable ($1 > E > 0$), and shrinking ($E < 0$) [19,20].

2.3.4. Kappa Coefficient

The Kappa coefficient is an index used for consistency testing proposed by Cohen in 1960 [21]. For images classification, consistency is whether the predicted results of the model are consistent with the actual classification results.

$$Kappa = \frac{P_o - P_c}{1 - P_c}, \tag{6}$$

In the formula, P_0 is the overall accuracy of the classification, which represents the probability that the classification result is consistent with the actual feature type for each random sample; P_c represents the probability that the classification result caused by chance is consistent with the actual feature type based on the evaluation criteria proposed by Cohen, higher than 0.8 can be regarded as the consistency of the best gradient. In this study, the calculation of the Kappa coefficient is based on the confusion matrix report from Envi 5.3.

3. Results

3.1. *The Dynamics of Fish Ponds across the GBA*

Figure 7 shows the overview of pond dynamics over the 10 periods in GBA. Based on the water area change index (*W*) in Figure 8, it can be seen that the total pond area firstly showed an incipiently increasing trend, followed by a decreasing trend. The fluctuation of W is basically consistent with the evolution of fish ponds in the GBA. From 1986 to 1994, fish ponds demonstrated a continuous rise from 106,603 hm^2 to 187,153 hm^2, which is an enlargement of with a total of area of 80,550 hm^2 or a relative increase of 75.5%. Accordingly, *W* also performed an accelerated expansion and peaked in 1994. Since then, the areal change fluctuated within a small range until 2009, which means that fish ponds were under a relatively stable status during that period. In 2009, the area fell slightly to 184,589 hm^2, accompanied by an increase, and reached the maximum value of 196,326 hm^2 in 2013. After that, there was a decreasing trend from 2013 to 2019, with an average annual shrinkage of 10,242 hm^2; meanwhile, *W* also maintained a negative growth. By 2019, the area had shrunk to 134,874 hm^2 with an overall decrease of 31.3%, but the total area was still larger than 1986.

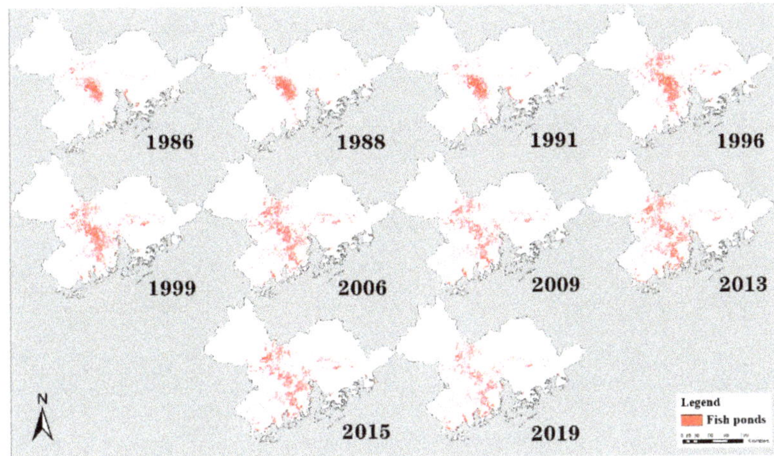

Figure 7. Spatial distribution of fish ponds in the GBA in the 10 different study periods.

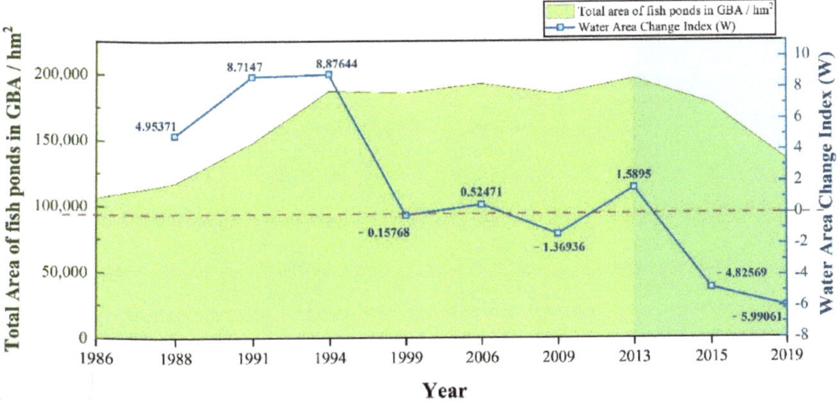

Figure 8. Water area change index and pond areal fluctuation in the GBA from 1986 to 2019.

According to the *SD* change for each period, the spatial distribution of fish ponds in the whole GBA could be divided into three phases (1986–1994, 1994–2013 and 2013–2019 in Figure 9, and Table 1). The increasing phase between 1986 and 1994 indicated that the differences in spatial distribution of the fish ponds increased during this phase. Geographically, it could also be seen in Fig. 4 that the fish ponds were mainly located in the middle of the GBA in early period, but gradually spread to the east and south of the GBA. The *SD* index dropped to 63,026.68 in 2009, showing fish ponds turned into a diffused spatial distribution. This can also be seen in Figure 7, the pond density increased, and the spatial distribution expanded outward. After they rebounded in 2013, the index kept decreasing to 52,474.643 in 2019, illustrating that the distribution shows a trend of initial agglomeration and subsequent diffusion between 2009 and 2019.

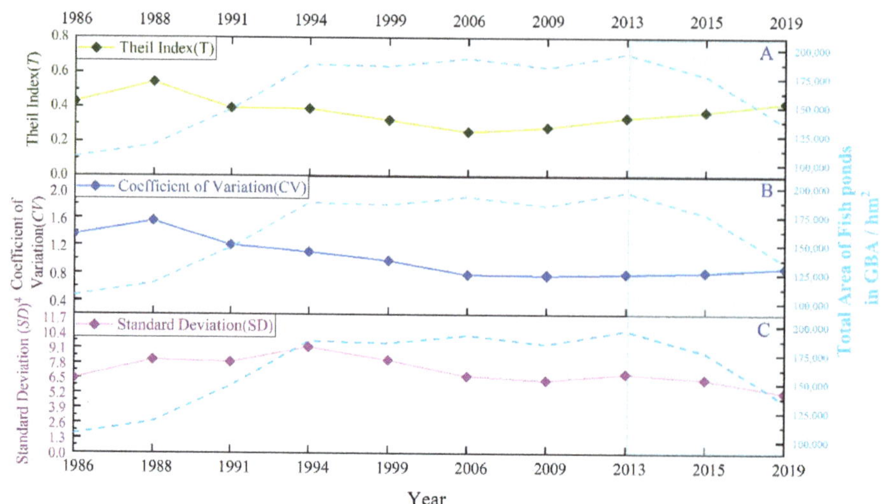

Figure 9. Fluctuation of three statistical indices compared to the areal changes of fish ponds from 1986 to 2019. (**A**): Theil index (*T*); (**B**): coefficient of variation (*CV*). (**C**): standard deviation (*SD*).

Table 1. Statistical results of fish ponds in GBA from 1986 to 2019.

Years	Area/hm^2	Standard Deviation	Coefficient of Variation (*CV*)	Theil Index (*T*)
1986	106,602.8	64,359.503	1.358	0.428
1988	117,164.4	80,927.218	1.554	0.544
1991	147,796.0	78,997.455	1.203	0.393
1994	187,153.1	91,933.360	1.105	0.389
1999	185,677.6	80,891.375	0.980	0.324
2006	192,497.4	66,908.650	0.782	0.257
2009	184,589.5	63,026.680	0.768	0.282
2013	1963.5.6	68,612.618	0.786	0.338
2015	177,377.5	63,978.322	0.812	0.373
2019	134,873.6	52,474.643	0.875	0.424

The coefficient of variation and Theil coefficient in each year reflected that, there were large differences in the size and area of fish ponds in the early stage, but the ponds became increasingly homogeneous in the later period. From 1986 to 2009, the coefficient of variation dropped significantly from 1.358 in 1986 to 0.768 in 2009, and the Theil coefficient dropped from 0.428 to 0.282, indicating that the differences in pond size gradually decreased during this period. In the next stage 2009–2019, the Theil coefficient fluctuated basically in a relatively stable status. the variation coefficient also changed consistently to the Theil coefficient. The results proved that human activities have continuously influenced the distribution and size of fish ponds in the past 40 years. The fish ponds had transformed from an early near-natural ponds with different sizes and a near-natural random distribution into an artificial distribution and an artificial shape.

3.2. The Areal Variation Trends in the Prefectural Cities

In 1986, Foshan had the largest number of fish ponds, followed by Zhongshan and Zhaoqing. However, by 2019, Jiangmen has become the largest, followed by Foshan and Zhaoqing. Considering the area of the fish ponds and the overall areal variations at the initial and the end of the study period, the prefectural cities can be roughly divided into the following three categories:

The growing cities (Figure 10) include Guangzhou, Huizhou, Zhuhai, Zhaoqing, Jiangmen, and Zhongshan. The trend of areal variation from 1986 to 2019 shows a fluctuating growing trend, but the peaks appeared at slightly different times. The year 2013 was the peak of Guangzhou, Zhuhai, Zhaoqing, and Jiangmen. As for Zhongshan, in addition to the peak in 1994, it also reached a second peak in 2013. All the five prefectural cities experienced a decline after 2013. Similarly, Huizhou began to decline after a slight rise in the 2013–2015 period.

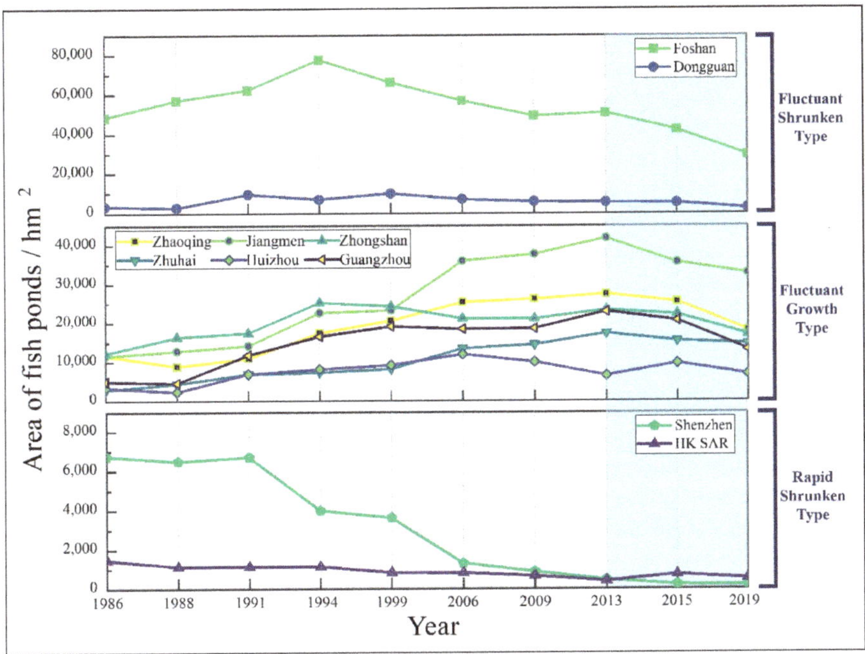

Figure 10. The trends for three representative changes in fish ponds in the GBA.

The shrinking fish ponds were observed in Foshan and Dongguan (Figure 10). In 2019, the area of fish ponds in Foshan decreased by 39.43%. In comparison, the area of fish ponds in Dongguan decreased by 27.90%. In terms of the variation trends, both cities maintained a fluctuated growth first and then a gradual shrinkage. The maximum area of the fish ponds in two cities happened in the 1990s. In 1994, Foshan's fish ponds reached the maximum area of 77,729 hm^2; in 1999, the area of Dongguan reached the highest value of 9995 hm^2. The minimums in both appeared in 2019, indicating that the shrinking trends will remain.

The rapid shrinking cities (Figure 10) include Shenzhen and Hong Kong. Overall, fish ponds in both cities have almost disappeared. By 2019, the area of fish ponds in Shenzhen decreased by 97.1%; the counterpart in Hong Kong also decreased by 63.8%. The areas of fish ponds in the two regions still maintained a decreasing trend with slight fluctuations, with the maximum value in 1986 and the minimum value in 2019.

It should be noted that, in 2013, except for Huizhou, Shenzhen, and Hong Kong, the other eight cities all reached a peak before showing a significant drop. Research on this milestone is meaningful to understand the development of fish ponds over the past 10 years.

3.3. The Spatio Patterns in Areal Dynamics among Study Units

The expansion coefficients (*E*) were calculated for the 25 study units in the GBA in each period. The study periods were divided into two stages–the first stage (1986–2013) and the second stage (2013–2019).

3.3.1. The First Stage (1986–2013)

At the first stage, the types of the study units were divided as follows:

Fast growing units, included Sihui, Sanshui, Panyu, Nansha, Xinhui, Taishan, Zhuhai, mainly located in central and southern GBA. Such cities obtained an expansion coefficient larger than 89.0, expansion rate greater than 5.6%, and internal structure transition coefficient, spatial structure transition coefficient both greater than 2.0. From 1986 to 2013, the total area of fish ponds in this units increased from 12,595 hm^2 to 80,369 hm^2, with a total of 538.10%, which made the areal proportion of these units increased from 11.81% to 40.94%.

Growing unites, included Gaoyao, Gaoming, Heshan, Nanhai, Zhongshan, and Combined units B, C, and E, mainly located in the western GBA. Such units received a expansion coefficient higher than 1.4, expansion rate greater than 0.8, and internal structure transition coefficient, spatial structure transition coefficient both greater than 0.3. From 1986 to 2013, the area of ponds in these units increased from 40,033 hm^2 to 77,009 hm^2, with a total of 92.36%, making the areal proportion of fish ponds in these units increased from 37.55% to 39.22%.

Stable unites, included five units in total, namely, Huizhou, Dongguan, Pengjiang, Jianghai, and Combined Unit B. Such units are relatively stable without significant increase: all of the expansion coefficient values were stable between 0 and 1, expansion rate and internal structure transition coefficient were both relatively low. During the first period, Dongguan maintained a relatively stable with slight growing, the apparent shrinkage of the area appeared in 2019, which lied in the second period, from 1986 to 2013. Although the area of fish ponds in this units increased from 12,937 hm^2 to 20,494 hm^2, a total of 58.41%, the pond areal proportion of the units decreased from 12.14% to 10.44%. Therefore, compared to other units, this increase was very insignificant.

Shrinking units, included Chancheng, Shunde, Shenzhen, Combined Unit D, and Hong Kong, mainly located in north-western and south-eastern GBA. These units obtained negative expansion coefficient and internal structure transition coefficient, with continued shrinking in total pond surface area. From 1986 to 2013, the total area of ponds in these units decreased from 41,038 hm^2 to 18,454 hm^2, a drop in 55.03%. The total pond areal proportion of the units decreased from 38.50% to only 9.40%.

3.3.2. The Second Stage (2013–2019)

Slight growing units, included only one unit of Taishan. From 2013 to 2019, the total area of ponds in this unit increased from 11,233 hm^2 to 13,430 hm^2, with a total of 19.5%. The relative areal proportion of this unit increased from 5.72% to 9.96%.

Stable unites, included Huizhou, Combined Unit E, and Hong Kong (Figure 11b). From 2013 to 2019, the total area of fish ponds in these units increased from 10,922 hm^2 to 11,725 hm^2, with a total of 7.35%. The relative areal proportion of these units increased from 5.56% to 8.69%.

Shrinking units, included Sanshui, Nanhai, Shunde, Gaoming, Chancheng, Panyu, Nansha, Dongguan, Shenzhen, Zhuhai, Zhongshan, Gaoyao, Sihui, Heshan, Xinhui, Jianghai, Pengjiang, Combined Units A, B, C and D, 21 units in all. It can be seen, during the stage, most of them were shrinking units. From 2013 to 2019, the area of fish ponds in these units dropped from 174,171 hm^2 to 109,719 hm^2, a total of −37.00%, making the relative areal proportion of this units decreased from 88.72% to 81.35%. At the same time, the water area change index (Figure 12) also showed that the average water body change index of each unit from 2013 to 2019 was −5.67%, and the average annual change for each unit was −4.11 hm^2. Among them, Shunde, Zhongshan and Sanshui, which were the top three in absolute annual shrinking in fish ponds, with an average annual shrinkage of 1164 hm^2,

1055 hm^2 and 969 hm^2, respectively. In Chancheng, Shenzhen and Heshan, where the average annual relative shrinking in fish ponds ranked the top three. The water body change index was −11.44%, −9.81% and −9.77% respectively.

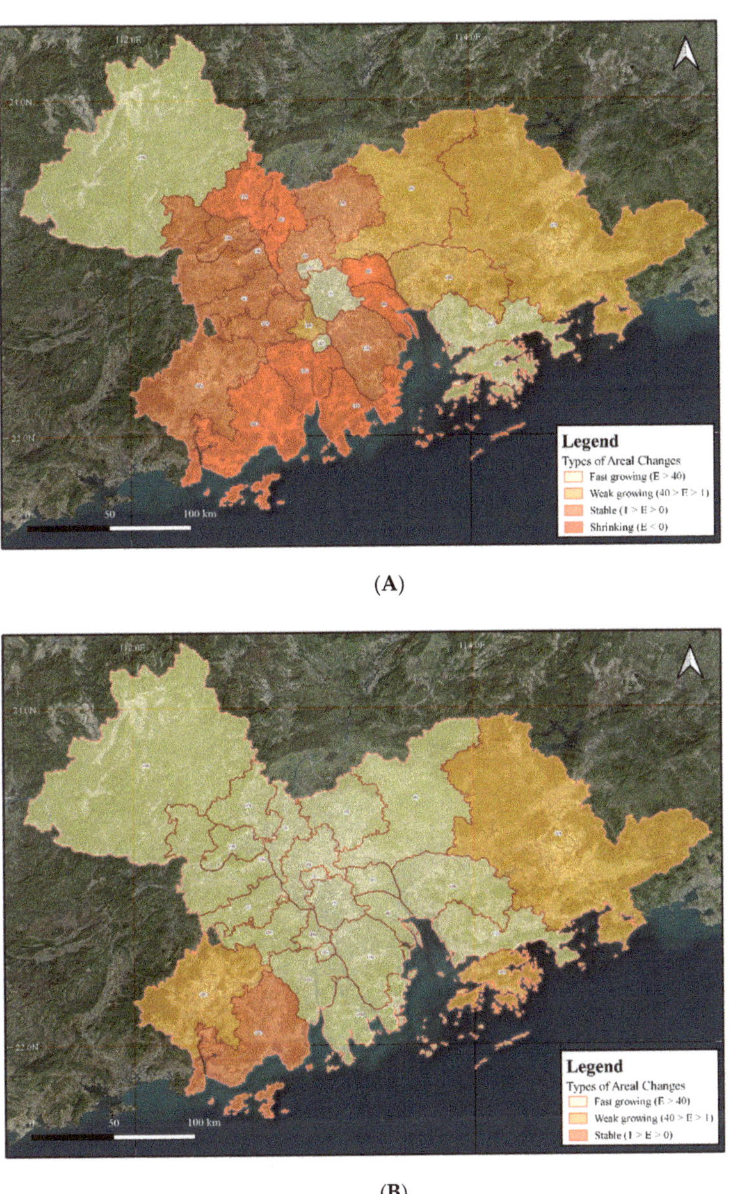

Figure 11. The spatiotemporal patterns in areal dynamics in the study units in the two different stages: (**A**) the first stage from 1986 to 2013; (**B**) the second stage from 2013 to 2019.

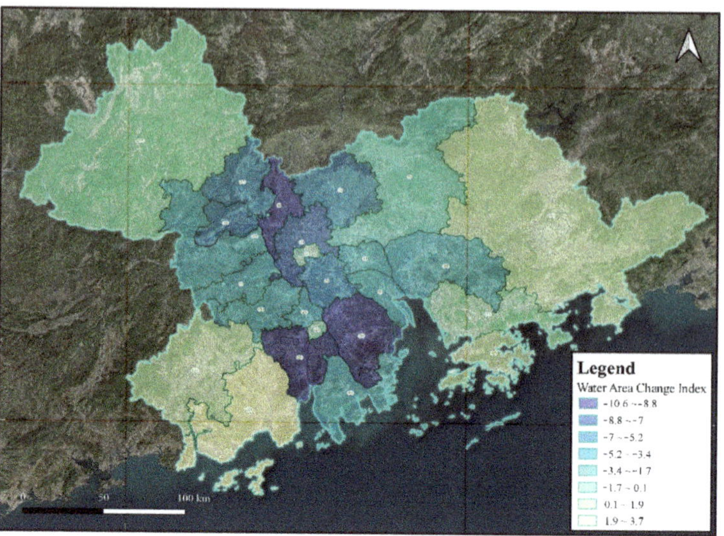

Figure 12. Distribution of each unit for waterbody area change index in GBA from 2013 to 2019.

It can be seen that taking 2013 as the milestone, the spatial dynamic patterns of fish ponds in the GBA have changed significantly. Except for the four units of Taishan, Hong Kong, Huizhou, Combined Unit B, which are classified as weakly expansionism or relatively stable, the other 21 units all experienced shrinking.

4. Discussion

4.1. Uncertainty Analysis

4.1.1. Accuracy Assessment

Previous studies have proved that the support vector machine (SVM) method has high interpretation accuracy in fish pond extraction [22]. This could also be seen in Figure 13A that the water body delineation using SVM was satisfactory.

Sometimes, problems happened to some narrow ponds with width less than 30 m, which could be false to be identified in Landsat imagery due to relative coarse spatial resolution (Figure 14) [23]. However, in consideration of the satellite return interval, Landsat satellites were the best choice. Table 2 has shown that the Kappa coefficient for each yearly classification results were all above 0.8, capable for the data analysis. To maintain the data quality, visual interpretation was combined to improve classification results; yet the result may still be affected by mixed pixels due to images' resolution (Figure 13B).

In addition, it should be noted that, because the study area is located in the subtropical area, there are almost no cloud-free satellite images available in the monsoon season. Most of the images used in this study were acquired during dry seasons. Generally, the surface area of the water bodies varies greatly during from monsoon season to the dry season, and the area during the dry season is commonly significantly smaller than in the monsoon season. However, as reported in Section 3, most of the fish ponds are now artificial, and changes in water surface area are rarely affected by season switch. Therefore, this impact can be safely ignored in this study. Besides, for a time scale of over 40 consecutive years, the Kappa coefficient of over 0.8 is well enough to perform a relatively accurate change trajectory of fish ponds in such a large scale of this study. However, it should be noted that, when similar methods are used to evaluate natural water bodies, this effect should be considered.

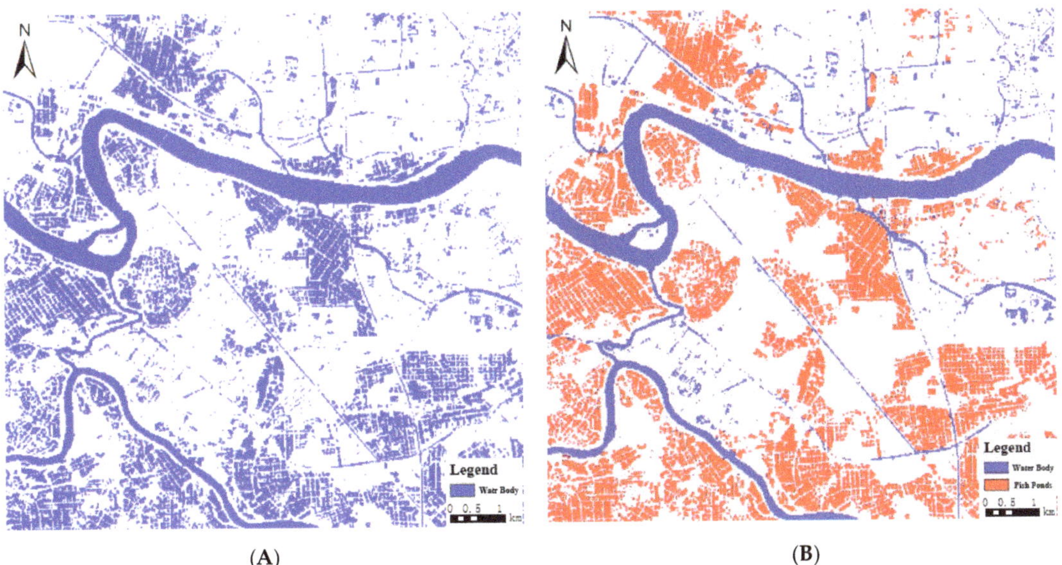

(A) (B)

Figure 13. Image classification results and the subsequent visual interpretation results: (**A**) the classification results using SVM; (**B**) the results of visual interpretation.

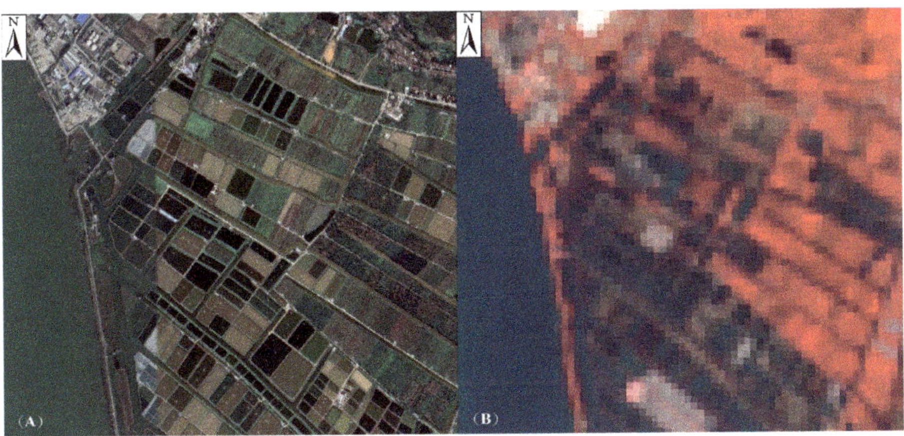

Figure 14. Fish ponds in Google Earth high resolution image (**A**) and Landsat image (**B**).

Table 2. Error Matrix of Classification Accuracy Assessment for Each Yearly Result.

Year	Kappa Coefficient	Year	Kappa Coefficient
1986	0.81	2006	0.86
1988	0.85	2009	0.83
1991	0.84	2013	0.80
1994	0.82	2015	0.81
1999	0.80	2019	0.80

4.1.2. Comparisons with Previous Studies

Previously, few studies were conducted on fish ponds in the entire Guangdong–Hong Kong–Macao Greater Bay Area. Many studies have been conducted in a specific city or a smaller area. Therefore, in the comparison, previous study results were compared with our results from the same specific areas. Previous studies about trends of dyke-ponds between 1978 and 2016 in Shunde, a district of Foshan City located in central GBA, reported that the spatial distribution of fish ponds in Shunde did not greatly change during 1988 and 1993 [24]. It was concluded from the comparison of fish pond distribution between these two years. Completing the gap of the time series by adding the fish ponds distribution in 1991 (Figure 15), we found that during this period, fish ponds in the eastern Shunde experienced a dramatic decrease, lost a large number of ponds, most of which disappeared in 1991.

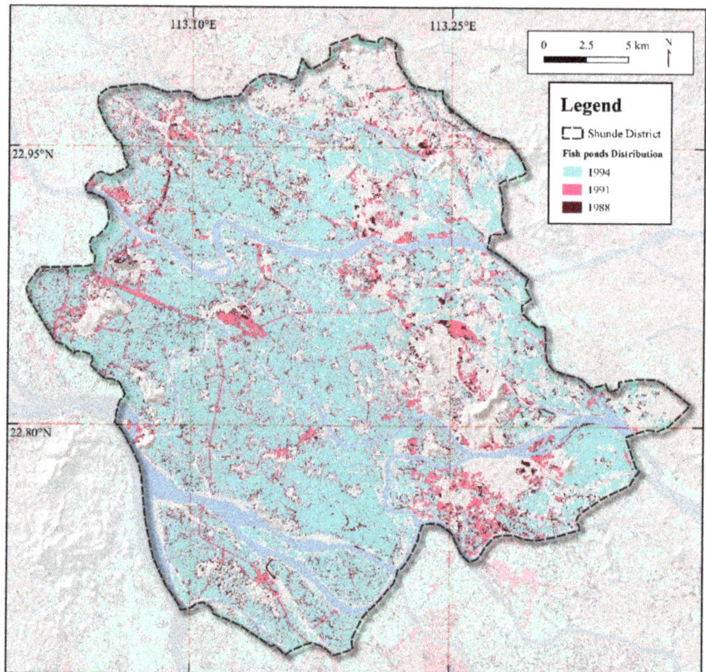

Figure 15. Spatial changes in fish ponds in Shunde District between 1988 and 1994.

Another previous study by Li [11] displayed the distribution changes of fish ponds in 1964, 1976, 1988, 2000, 2012 in Pearl River Delta region. The results indicated that fish ponds kept increasing in its area from 1964 to 2012 [11]. This study also covered the period of 1988 to 2012. However, due to some cities' boundaries adopted by the two studies are slightly different, resulting in false comparisons for such cities. The boundary of Guangzhou that two studies adopted are relatively consistent. Li's study displayed that the classification results of 1988, 2000, 2012, were 41.17 km^2, 203.06 km^2 and 213.44 km^2, respectively. Comparing with this study's results in 1988, and 1999 and 2013 (no results available for the years of 1999 and 2012), which are 45.86 km^2, 190.56 km^2 and 228.62 km^2, respectively, the two results are basically consistent, with an averaged relative difference of 10.93%. However, it should also be known that this difference may also be caused by the real difference in the latter two periods (1999 vs. 2000, 2012 vs. 2013). Despite the difference, the time series in our study more thoroughly demonstrate the trend of fish pond changes in the study area between 1988 and 2013 (Table 3). In addition, the obtained trends are quite consistent, but Li's study used merely 3-year data over a 24-year period, resulting in a harder response in sensitivity. This study revealed that fish ponds in Guangzhou actually experienced a fluctuation instead of a smooth increase, especially in the period of 1994 to 2013. Besides, the expansion of fish ponds happened before 2012.

Table 3. Comparison of the pond changes between Li's study and this study in Guangzhou for the period 1988 to 2013.

Year	Li's Study (km^2)	Year	This Study (km^2)	Percentage of Difference
1988	41.17	-	45.86	4.69%
-	-	1991	117.62	-
-	-	1994	165.15	-
2000	203.06	1999	190.56	12.49%
-	-	2006	184.20	-
-	-	2009	185.51	-
2012	213.44	2013	228.62	15.18%

4.2. Possible Causes for Pond Changes in Different Cities and Future Speculation

4.2.1. The Growing Cities

Guangzhou, Huizhou, Zhuhai, Zhongshan, Zhaoqing, Jiangmen—six cities in total were classified as growing cities. The reason was that compared with 1986, all of these cities experienced an increase in pond area in 2019. The peaks also appeared before 2019, showing a fluctuating growing trend.

Guangzhou's population increased by 1.67 million between 1990 and 2006, about 28% in total, yet the actual product consumption was about 3.7 times of the previous amount. The aquatic products in 2006 was 4.9 times of that in 1990 [22]. The inflow of population and the expanded market stimulated the demands and development of aquaculture, resultant increase in fish ponds in Guangzhou before the beginning of the new century, especially in the Nansha District of Guangzhou. Therefore, Nansha District experienced a strong expansion from 1986 to 2013. After 2013, Nansha District was designated as a free trade zone, and Nansha Port was also developed as a manufacturing and industrial export, resulting in further changes in land uses and shrinking in fish ponds [25]. Huizhou was similar too. The area of fish ponds near the seaside also maintained a positive growth before the development of the petrochemical industry before 2006, and then gradually shrank with industrial development [26].

Zhaoqing City also achieved rapid growth in fish ponds from 1986 to 2006 through the reconstruction of low-lying sandy wasteland [27]. Subsequently, from 2006 to 2009 and 2013 to 2019, due to the impact of natural disasters, bacterial diseases, and dramatic price fluctuation in aquatic products, the risk of aquacultural development increased, which dampened the enthusiasm of farmers for aquacultural development, and the area of fish ponds showed a rapid downward trend.

Driven by the adjustment of the regional agricultural policy, the area of fish ponds in Zhuhai had increased significantly before 2006. During the period 1990–2006, local cultivated land experienced an accelerated loss, during which the net transfer area of cultivated land to the fish ponds reached 16,054.23 hm^2 [28,29]. Subsequently, urbanization developed during 2006–2009. As there were less cultivated land and forest land available for urbanization, the occupation of fish ponds was accelerated to a certain extent. The Government planned to accelerate the development of ecological fisheries and the construction of agricultural and fishery infrastructure to provide a guarantee for the development of fish ponds. Similarly, the Local Government of Jiangmen released a new policy [30] in 2009 to prompt aquacultural development the growth of the fish ponds in Jiangmen from 2009 to 2013.

However, from 2013 to 2019, Zhuhai has invested heavily in the development of ecological agriculture such as flowers, fruits and vegetables, and organic rice. The Government introduced modern agricultural industrial parks, such as the Yongcheng Horticulture and Taiwan Orchid Greenhouse Planting Base, and eliminated fish ponds with low economic and environmental benefits, causing the shrinking of fish ponds.

4.2.2. The Shrinking Cities

The fish ponds in Foshan and Dongguan showed an initial increasing trend followed by a significant shrinking. From 1986 to 1994, the area of fish ponds in Foshan City has increased significantly. Taking the Nanhai County in Foshan City as an example, the county became a pilot for land policy reform in 1987. Stimulated by the flexible land policy, the productivity has greatly increased. The traditional agriculture was drastically reduced and transformed into vegetable planting and aquaculture. After 1994, Nanhai county was cancelled and became a district of Foshan. In order to promote industrialization, the Government reduced the rent of collective land in rural areas to attract a large number of enterprises to settle in. The upgradation in the industry prompted a rapid expansion of built-up land, with an average annual growth rate of over 7% [31]. In addition, township and village enterprises had sprung up all over Foshan. In 1991, the total income of township and village enterprises in Foshan reached 21.04 billion Chinese yuan, which increased by 3.84 times in 1997 [32]. The expansion of built-up land encroached on a large amount of fish ponds, resulting in shrinkage of fish ponds and its fragmented distribution. However, in the most recent 5 years, a series of development policies, such as the "Conservation and Development Plan for the Agricultural System of Fish Ponds in the Pearl River Delta, Foshan, Guangdong" and "Strategic Plan for the Implementation of Rural Revitalization in Guangdong Province (2018–2022)" [33] have slowed down the shrinking. At the same time, the Foshan Local Government has declared the fish ponds as an important agricultural cultural heritage in China, and combined it with the tertiary industry, which had a certain effect on protecting the fish ponds and increasing their outputs. Such measures have slowed down the shrinking trend from 2013 to 2019, but remained insufficient in reversing the shrinking trend.

4.2.3. The Fast Shrinking Cities

The number of fish ponds in Hong Kong SAR and Shenzhen was relatively small, but they have shrunk at a rapid rate (Figure 16). The fish ponds in Hong Kong are mainly located on the river alluvial plains in the estuary of the Shan Pui River in Yuen Long and Nan Sang Wai. In the 1980s, the development of fish ponds in Hong Kong reached its peak. However, with the rapid development of Yuen Long after the 1990s, the land used for the fish ponds in Nan Sang Wai had changed. For example, the original fish ponds were excavated and converted to a new channel of the Kam Tin River or recreated as smaller triangular ponds near the river channel. Moreover, the discharge of wastes in the process of urbanization had led to the continuous deterioration of the water quality of Shan Pui River, which is not suitable for aquaculture now, resulting in the disposal of local fish ponds.

The fish ponds in Shenzhen are mainly concentrated in the original Bao'an County. After the establishment of the Shenzhen Special Economic Zone in 1980, superior policies and other geographical conditions promoted the rapid development of industry. The demand for built-up surged, and changes in land uses led to a sharp decline in fish ponds over the past 40 years. Therefore, the urbanization is the major cause for fish pond shrinking in Shenzhen and Hong Kong. Besides, large amounts of aquatic products imported from mainland China was also an important cause for the shrinking of fish ponds in Hong Kong.

Figure 16. The rapid shrinkage of fish ponds in Shenzhen and Hong Kong SAR in the 1990s.

4.2.4. Future Trends of Fish Ponds in GBA

In view of the current trends of changes in various cities and the overall planning background of the Greater Bay Area, the shrinkage of fish ponds in the future will remain for a long time. However, the rate of shrinkage in various regions will vary greatly due to various drivers such as local development policies. It is expected that the development of fish ponds will tend to integrate with the tourism and service industry. The economic benefits of fish ponds will increase accordingly via excavation of the cultural value of fish ponds and construction of traditional aquacultural demonstration areas. Due to the differences in regional development and local government investment, the better evolution in fish ponds could appear in such regions where the economic development is quite high.

5. Conclusions

In this study, we analyzed spatiotemporal changes in mulberry-dyke-fish ponds in the GBA using Landsat satellite images obtained from 1986 to 2019. combined the measurements of standard deviation, coefficient of variation, Theil coefficient, water body change index as well as expansion coefficient, the spatiotemporal changes were quantified.

From 1986 to 2019, the fish ponds in the GBA showed an overall increasing trend in the first stage and a sharp decreasing trend at the second stage. The year of 2013 was a milestone. A total 25 study units in the two periods before and after 2013 were studied. It was found that 18 units were transformed into shrinking trend. Additionally, the causes for the fish pond changes were analyzed, and the future development of fish ponds was also predicted. The results proved that human activities have continuously influenced the spatial distribution and size of fish ponds in the past 40 years. The fish ponds had transformed from a near-natural ponds with different sizes and a near-natural random distribution in the early stage into an artificial distribution and an artificial shape.

The increased social demands were the major causes for the steady growth in fish ponds in the GBA from 1986 to the beginning of the 21st century. The market price of aquatic products directly affected farmers' willingness to develop and maintain fish ponds. The policies reflected the local governments' attitudes toward the development of fish ponds. Urbanization was the main cause of shrinkage in fish ponds. The shrinkage of Shenzhen and Hong Kong before 2013 was due to the encroachment of urban expansion, while Foshan was due to the introduction of a large number of enterprises to promote industrial upgradation. After 2013, the policy for the development of metropolis in the GBA was given priority, and the development tended to be economic development and industrial upgradation. the outputs of aquacultural ponds are relatively low, thus, the shrinking of fish ponds will remain in future. However, increasing the economic outputs of fish ponds through deep excavation of the cultural value and construction of some new aquacultural demonstration areas can slow down the shrinking trend and enhance the social and cultural values of fish ponds.

Author Contributions: Investigation, Z.C., X.Y., W.Z. and J.Q. data preparation, Z.C. and W.Z. writing—original draft preparation, W.Z. and J.Q. writing—review and editing, X.Y., E.P., L.R. and X.X. supervision, X.Y. project administration, X.Y., L.R. and X.X. Funding acquisition, X.Y. and X.X. All authors have read and agreed to the published version of the manuscript.

Funding: The National Natural Science Foundation of China (Grant No.: 41871017 and 41476152) funded this research.

Institutional Review Board Statement: Not applicable.

Informed Consent Statement: Not applicable.

Data Availability Statement: Not applicable.

Acknowledgments: The authors thank the anonymous reviewers for their careful reading of the manuscript and their many insightful comments and suggestions.

Conflicts of Interest: The authors declare no conflict of interest.

References

1. Liu, S.; Min, Q.; Jiao, W.; Liu, C.; Yin, J. Integrated emergy and economic evaluation of Huzhou mulberry-dyke and fish-pond systems. *Sustainability* **2018**, *10*, 3860. [CrossRef]
2. Zhong, G. The characteristics of dike-pond system and the practical significance. *Sci. Geogr. Sin.* **1988**, *8*, 12–17, 99.
3. Astudillo, M.F.; Thalwitz, G.; Vollrath, F. Modern analysis of an ancient integrated farming arrangement: Life cycle assessment of a mulberry dyke and pond system. *Int. J. Life Cycle Assess.* **2015**, *20*, 1387–1398. [CrossRef]
4. Nie, C.; Li, H. The integrated Dike-pond systems: Its new developments and ecological problems. *J. Foshan Univ.* **2001**, *19*, 49–53.
5. Chen, W.; He, B.; Nover, D.; Lu, H.; Liu, J.; Sun, W.; Chen, W. Farm ponds in southern China: Challenges and solutions for conserving a neglected wetland ecosystem. *Sci. Total Environ.* **2019**, *659*, 1322–1334. [CrossRef]
6. Li, M.; Zhang, Y.; Xu, M.; He, L.; Liu, L.; Tang, Q. China Eco-Wisdom: A Review of Sustainability of Agricultural Heritage Systems on Aquatic-Ecological Conservation. *Sustainability* **2020**, *12*, 60. [CrossRef]
7. Guo, S. The Value, Protection and Utilization of Mulberry Ponds in the Pearl River Delta from the Perspective of Agricultural Cultural Heritage. *Trop. Geogr.* **2010**, *30*, 114–120.
8. Liu, K.; Xie, L.; Zhuang, J. Analysis on the evolution of the spatial pattern of the dyke-pond system in Foshan city. *Trop. Geogr.* **2008**, *28*, 513–517.
9. Liu, Z.; Cui, W. Multi-temporal Remote Sensing Analysis of Foshan Dyke-Pond System. *Guangdong Agric. Sci.* **2017**, *44*, 126–130, 183.
10. Lin, M.; Ji, S. Analysis on mode change and landscape pattern of the Dike-pond agriculture in Zhongshan. *Guangdong Agric. Sci.* **2014**, *41*, 192–197, 205, 245.
11. Li, Y.; Liu, K.; Liu, Y.; Zhu, Y. The Dynamic of Dike-Pond System in the Pearl River Delta during 1964–2012. In *Global Changes and Natural Disaster Management: Geo-Information Technologies*; Springer: Berlin/Heidelberg, Germany, 2017; pp. 47–59.
12. Mao, Y. The Theoretical Basis and System Innovation of the Construction of Guangdong-Hong Kong-Macao Greater Bay Area. *J. Sun-Yatsen Univ.* **2019**, *59*, 173–182.
13. Han, X.L.; Yu, K.J.; Li, D.H.; Wang, S.S. Building the Landscape Security Pattern of Dike-Pond System with Urban Functions A Case of Magang Part of Shunde District, Foshan City. *Areal Res. Dev.* **2008**, *123*, 109–112, 130.

14. Yan, Y.; Li, Y. The Comparative Study of Remote Sensing Image Supervised Classification Methods Based on ENVI. *Beijing Surv. Mapp.* **2011**, *11*, 14–16.
15. Gong, Z. Ponderation over the Flood Control Functions of the Artifical Landforms in PRD. *Trop. Geogr.* **2012**, *32*, 50–56.
16. Wei, C.T.; Hannes, B.T. Measuring urban agglomeration using a city-scale dasymetric population map: A study in the Pearl River Delta, China. *Habitat Int.* **2017**, *59*, 32–43. [CrossRef]
17. Yang, X.; Lu, X. Drastic change in China's lakes and reservoirs over the past decades. *Sci. Rep.* **2014**, *4*, 1–10. [CrossRef] [PubMed]
18. Hui, L. Regional inequality measurement: Methods and evaluations. *Geogr. Res.* **2006**, *45*, 152–160.
19. Ye, C. Change characteristics and spatial types of dike-pond in the Pearl River Delta. *J. East. China Inst. Technol.* **2013**, *12*, 315–322.
20. Zhang, W.; Xue, D. Urbanization base of city-land use expansion in the Zhujiang River Delta. *J. Nat. Resour.* **2003**, *18*, 64–71.
21. Cohen, J. A coefficient of agreement for nominal scales. *Educ. Psychol. Meas.* **1960**, *20*, 37–46. [CrossRef]
22. Wu, Y. Analysis of the indicators between urban metabolism and land use change in Guangzhou. *Geogr. Res.* **2011**, *14*, 478.
23. Yang, X.; Lu, X. Delineation of lakes and reservoirs in large river basins: An example of the Yangtze River Basin, China. *Geomorphology* **2013**, *190*, 92–102. [CrossRef]
24. Li, F.; Liu, K.; Tang, H.; Liu, L.; Liu, H. Analyzing Trends of Dike-Ponds between 1978 and 2016 Using Multi-Source Remote Sensing Images in Shunde District of South China. *Sustainability* **2018**, *10*, 3504. [CrossRef]
25. Li, J.; Du, Y.; Cai, A. Change Characteristics of Coastal Wetlands in the Pearl River Delta under Rapid Urbanization. *Wetl. Sci.* **2019**, *17*, 267.
26. Chen, J. *The Evolution of Reclamation Area and Its Impact on Land Resource Utilization*; Guangzhou University: Guangzhou, China, 2018.
27. Yang, M.; Qu, L.; Zhang, Y. The Driving Mechanism of Cultivated Land Change in the Development and Utilization of Core Regions in the Outer Peripheral of the Pearl River Delta in the Past Ten Years—A Case Study of Dinghu District. In Proceedings of the Annual Conference of China Land Society 2008, Hefei, China, 12 October 2008; pp. 599–603.
28. Wu, D.; Dong, Y.; Liu, H.; Ni, S. Analysis on the Spatial-temporal Changes of Cultivated Land and Their Driving Forces in Zhuhai. *Trop. Geogr.* **2009**, *29*, 472–476, 482.
29. Zhong, S. *Reasearch on Coordinated Development of Regional Resources, Environment and Economy—A Case Study of Zhuhai City*; Jilin University: Changchun, China, 2013.
30. Zhuhai Municipal Bureau of Agriculture. Work Plan for the Construction of Jiangmen Modern Agriculture Comprehensive Demonstration Base. 2009.
31. Liu, Y.; Lin, H. Spatial-Temporal Changes and Mechanism of Land Use in the Pearl River Delta from the Perspective of Property Rights Alteration: A Case Study of Nanhai District in Foshan City. *Trop. Geogr.* **2019**, *39*, 54–65.
32. Gui, D. A Preliminary Study on the Development Track of Foshan's Private Economy. *Guangdong Econ.* **2010**, *162*, 47–53.
33. Government of Guangdong Province. Strategic Plan for the Implementation of Rural Revitalization in Guangdong Province (2018–2022). 2019.

MDPI

Article

Improving Transboundary Drought and Scarcity Management in the Iberian Peninsula through the Definition of Common Indicators: The Case of the Minho-Lima River Basin District

Rodrigo Maia [1,2,*], Miguel Costa [1,2] and Juliana Mendes [1,2]

1 Civil Engineering Department, Hydraulics, Water Resources and Environment Division, Faculty of Engineering, University of Porto, Rua Roberto Frias, 4200-465 Porto, Portugal; mapcosta@fe.up.pt (M.C.); juliana@fe.up.pt (J.M.)

2 Interdisciplinary Centre of Marine and Environmental Research (CIIMAR), University of Porto, Terminal de Cruzeiros do Porto de Leixões, Avenida General Norton de Matos, 4450-208 Matosinhos, Portugal

* Correspondence: rmaia@fe.up.pt

Citation: Maia, R.; Costa, M.; Mendes, J. Improving Transboundary Drought and Scarcity Management in the Iberian Peninsula through the Definition of Common Indicators: The Case of the Minho-Lima River Basin District. *Water* **2022**, *14*, 425. https://doi.org/10.3390/w14030425

Academic Editors: Athanasios Loukas and Luis Garrote

Received: 31 December 2021
Accepted: 27 January 2022
Published: 29 January 2022

Publisher's Note: MDPI stays neutral with regard to jurisdictional claims in published maps and institutional affiliations.

Copyright: © 2022 by the authors. Licensee MDPI, Basel, Switzerland. This article is an open access article distributed under the terms and conditions of the Creative Commons Attribution (CC BY) license (https://creativecommons.org/licenses/by/4.0/).

Abstract: Drought is one of the most damaging natural hazards in the Iberian Peninsula, causing varied socioeconomic and environmental impacts. To prevent these impacts, there must be close cooperation between Portugal and Spain, as the two countries share five river basins. However, regarding drought planning and management the two countries are clearly in different stages. Portugal approved a national drought plan in 2017, while Spain has already had drought plans in place for all River Basin Districts since 2007 and approved an updated version of these plans in 2018. The Spanish drought plans currently in place foresee two sets of indicators: prolonged drought and scarcity indicators. This paper presents the definition of similar indicators for the Portuguese part of the shared Minho and Lima river basins, according to European guidelines and in common with Spain, with the aim of developing a joint international drought management plan for these basins. For the period from October 1980 to September 2017, the comparison of the indicators obtained for the Portuguese parts of the basins with the corresponding Spanish ones shows a similarity in the occurrence of drought and scarcity in both parts of the basins, although with a higher prevalence of scarcity situations in the Lima Spanish part. This work was developed in close collaboration with the River Basin District competent authorities of both countries, aiming to be a prototype for the definition of new and comparable drought and operational scarcity indicators. Therefore, this work is a starting point for the creation of common tools for integrated drought management of transboundary basins in the IP.

Keywords: prolonged drought; scarcity; indicators; transboundary river basins; Iberian Peninsula

1. Introduction

Drought and water scarcity have always been situations of concern. While drought is a natural phenomenon caused by an abnormal precipitation deficit over a certain region and period of time, water scarcity is a result of human action, referring to an insufficient water availability to satisfy water demands for different socio-economic uses [1]. From an operational perspective (as mostly considered for definition of common indicators), scarcity is considered a temporal problem of lack of water resources. Nonetheless, water scarcity can also be understood as the long-term unsustainable use of water resources, determined by social and political processes [2,3].

Drought and water scarcity events commonly coexist and are interdependent in a same region, making the distinction between them a complex process. Moreover, water scarcity enhances regional vulnerability to drought effects [3,4].

The increasing frequency and intensity of such phenomena, which are expected to worsen from climate change, have been leading to an increasing and urgent need to establish specific policies and management measures to deal with the associated risks and the mitigation of the related effects [5,6]. These issues become even more relevant in watersheds shared by different countries, since conflicts (defined by opposing national interests and/or policies) over water use tend to be more pronounced, and the harmonization of water policies and bilateral agreements and sustainable water use is required.

At the European Union (EU) level, in order to achieve the environmental objectives established for the protection of water bodies, the Water Framework Directive (WFD) [7] states that, within a river basin shared by different countries, where the use of water may have transboundary effects, water management should be coordinated for the whole of the River Basin District (RBD). In this context, the WFD stipulates that Member States (MSs) shall ensure coordination with the aim of producing a single international River Basin Management Plan (RBMP). In case that is not possible, the WFD envisages that the MSs develop individual but coordinated RBMPs for each territorial part of the river basin. This directive corresponds to greater country responsibility and interference in water management of shared river basins in order to achieve WFD provisions of good environmental status, in line with increasing concern with water quality, namely by southern Member States [8].

One of the purposes of the WFD is to contribute to the mitigation of drought effects. However, the WFD states that in the case of exceptional natural conditions, such as a prolonged drought, the temporary deterioration of water quality will not breach the Directive requirements (Article 4.6). Although "water scarcity" is not directly mentioned in the WFD, its frame is implicit across it, namely by highlighting (Article 11) that MSs shall ensure the establishment of measures to promote efficient and sustainable water use.

To deal with more specific situations, namely for the mitigation of drought and scarcity effects, the WFD highlights the possibility of complementing the RBMPs through special programs and management plans (e.g., drought management plans—DMPs), which should, whenever possible, also be jointly set by the States involved in the case of international river basins (Article 13.5). Despite not being mandatory, the development of DMPs has been strongly recommended, namely by means of the EU Communications on Drought and Water Scarcity [9–11] and corresponding follow-up reports. In one of those reports [12] the importance and the need of separate indicators to evaluate drought and water scarcity situations is enhanced.

In the Iberian Peninsula (IP), drought is one of the most damaging natural hazards, causing wide-ranging socioeconomic and environmental impacts, which are prospected to continue the recent years' aggravation trends due to climate change. To prevent impacts from drought, close cooperation between Portugal and Spain is required, namely regarding water and drought planning and management, as the two countries share five river basins (Minho/Miño, Lima/Limia, Douro/Duero, Tejo/Tajo and Guadiana) that cover 45% of the Iberian territory (Figure 1). This issue is particularly relevant for Portugal, as 64% of its territory corresponds to shared river basins, with the Portuguese part located downstream, rendering the country extremely vulnerable to the quantity and quality of water flowing from Spain [13].

In 1998, the two countries signed the Convention on Cooperation for Portuguese-Spanish River Basin Protection and Sustainable Use (commonly referred to as the Albufeira Convention), which has been in force since 2000 [14] and was revised in 2008 [15] under the WFD principles. The convention defines the framework for bilateral cooperation to promote and protect the good status and the sustainable use of water resources in the shared river basins, as well as actions to contribute to mitigate the effects of floods, droughts and water scarcity situations.

Under the Albufeira Convention, two institutional bodies were constituted: the Parts Conference (Article 21) and the Commission for Convention Development and Appliance (CADC) (Article 22). The first has mainly a political role and guarantees bilateral cooperation at the highest levels, its representatives being appointed by the Government of each country. The CADC has an operational, deliberative, consultative and supervisory role, having the responsibility to ensure compliance with the Convention obligations. The CADC is composed of two delegations, one from each country [14].

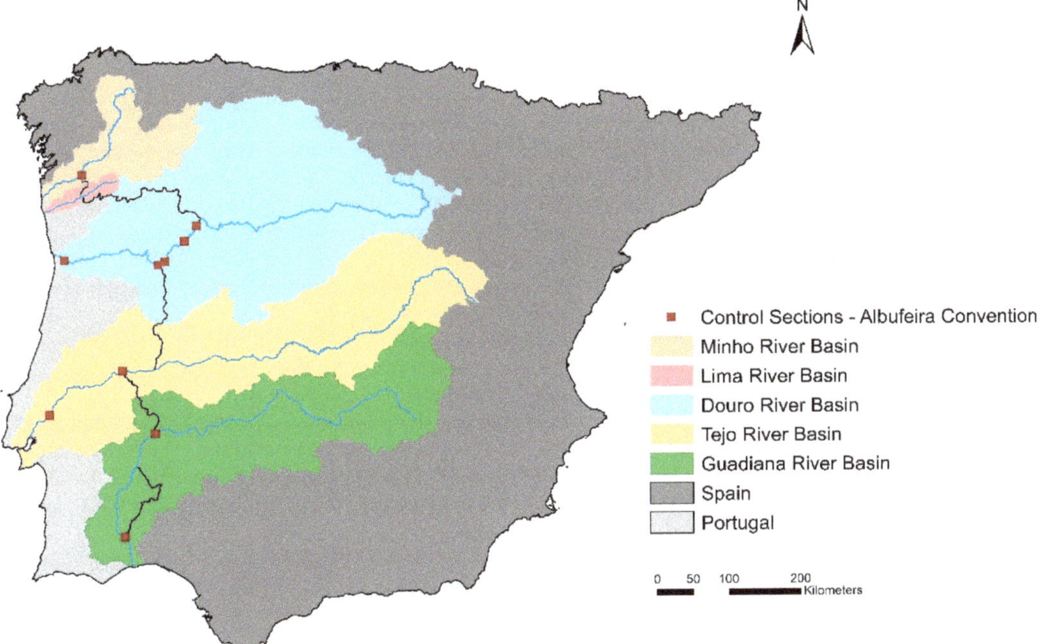

Figure 1. Transboundary river basins in the Iberian Peninsula and respective flow control sections under the Albufeira Convention.

In this context, and in order to secure good water conditions and the current and predictable uses, the Albufeira Convention (AC) defined a minimum flow regime (MFR) at the border sections (and bordering/international river stretches) of each of the Portuguese-Spanish shared river basins (except Lima) (Figure 1). Currently, as stated in the revised version of the convention [15], the MFR consists of minimum volumes of water to be guaranteed at the control (namely bordering) sections: annually and quarterly, for all (4) shared basins; weekly for the two major basins (Douro/Duero and Tejo/Tajo rivers), and; for the southern basin (Guadiana river basin), mean diary flow at the bordering section at the entrance of the Guadiana river in Portugal (Badajoz weir) and at the upstream section (Pomarão) of the estuarine and lower bordering stretch between the two countries. Nevertheless, these flow regimes are not applicable under exceptional drought conditions, those defined by means of cumulative precipitation thresholds and also (only for Guadiana river) on reservoir volumes, based on the weighted values of the referenced monitoring stations in the Spanish part of each basin [14,15].

Moreover, according to Article 19 of the Convention, the parties: (i) should coordinate their actions to control and to prevent drought and scarcity situations and (ii) should define the nature of the exceptions and the establishment of exceptional mechanisms, which can include, among others: (a) the definition of the conditions in which the exceptional measures can be applied and (b) the possible use of indicators that characterize, in an objective way, the drought and scarcity situations. Under the above-referred context and Albufeira Convention principles, the current MFR values and time frame continue to be considered provisional, insufficient and requiring of further revision [16,17].

Spain approved and has implemented drought plans in all the River Basin Districts since 2007 [18]. The 2007 Spanish drought plans were already been revised and approved in

2018 [19]. In the 2018 versions (e.g., [20]), two types of indicators are defined: a Prolonged Drought Indicator (PDI) and a Water Scarcity Indicator (SI).

Portugal approved a national drought plan in 2017, by which two types of drought situations (agrometeorological and hydrological) are identified through the use of simple variables (or standardized indexes), like precipitation, stored dam reservoir volumes, piezometric levels and soil moisture [21]. Specific DMPs by River Basin District are envisaged but still not active.

Considering these national differences, in order to minimize water conflicts and prevent drought-related impacts (e.g., in 1994/95, in the Guadiana river basin, the affluences to Portugal were actually null for six months [22]. In 2019, some Portuguese parts of the Tejo river basin dried out due to Spanish water management, contributing to the water quality degradation in the Portuguese part [23], which led to political contact between the two Government representatives, namely the Environmental Ministers [24]), several efforts should be made by the two countries to adopt coordinated and/or possible joint drought management and planning through the establishment of standard and/or common approaches, in the light of the AC and WFD. One of the most important approaches that should be enforced is the definition of common indicators that characterize, in an objective way, the drought and scarcity situations [13,22,25].

In this sense, the Spanish methodology, by using separate indicators for those situations, presents a good basis for the establishment of a common system of indicators for the Iberian Peninsula in order to achieve a better integration with the WFD and AC goals, representing a step forward in the implementation of European Water Policies regarding drought management and planning.

The purpose of the presented work was the definition of prolonged drought and scarcity indicators for the Portuguese parts of the shared Minho and Lima river basins, in common with the Spanish parts of the basins, as the basis for the joint international DMP for these basins. In fact, following this work, this DMP is already in the stage of discussion for approval by the Portuguese and Spanish RBD authorities (respectively, APA—Agência Portuguesa do Ambiente and CHMS—Confederación Hidrográfica del Miño-Sil).

The paper describes the procedure used in the definition of the indicators for the Portuguese parts of the basins, as well as its comparison with the corresponding Spanish indicators, for a period from October 1980 to September 2017. The work allowed us to assess the applicability of a common system of indicators for any Portuguese-Spanish transboundary river basin's drought and scarcity management.

This work is intended to be a prototype for the definition of new and similar drought indicators to be applied in common by Portugal and Spain in the shared river basins, and was developed in close collaboration with the Portuguese Minho and Lima RBD and the corresponding Spanish RBD authorities under the scope of the RISC-ML project [26].

2. Materials and Methods

2.1. Study Area

The study area encompasses the Minho and Lima river basins, which are two of the five international river basins shared between Portugal and Spain that are integrated into a single international RBD, in accordance with the WFD. The major part of the RBD total area is in Spain, and only about 10% of that area is located in Portugal, that corresponding to 48% of Lima and 5% of Minho total river basins area (respectively, 2522 km^2 and 17,067 km^2) [27].

Regarding water management, in Portugal, the Minho and Lima river basins (Portuguese parts) integrate and form the Minho-Lima RBD (RBD1), with each river basin part being considered a specific and independent hydrological analysis unit. In Spain, the two river basins' Spanish parts constitute the Miño-Sil RBD, which is divided in 6 territorial operational management units (TMUs) (5 in Minho, 1 in Lima), as shown in Figure 2.

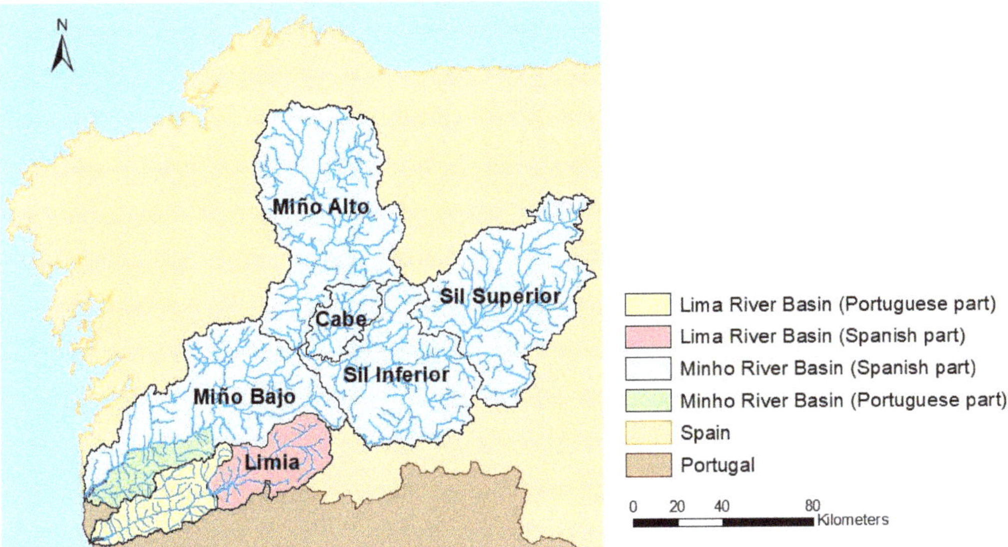

Figure 2. Portuguese and Spanish parts of the Minho and Lima river basins, with their corresponding Spanish territorial management units (TMUs) identified.

2.2. Methodology

In this work, the definition of the Prolonged Drought and Scarcity indicators for the Portuguese part of Minho and Lima river basins was carried out according to the methodology defined in the current version of the Special Drought Plan of the Spanish part of the Miño-Sil River Basin District (2018 PES-MS) [20], in agreement with the technical instruction provided by the Ministry for the Ecological Transition of Spain [28].

A prolonged drought is a natural, persistent and intense situation of reduction of precipitation, produced by unusual circumstances, with influence on the runoff. The Prolonged Drought Indicator (PDI) should identify, temporally and territorially, runoff reduction by natural causes, independently of human water resources management. Therefore, the aim of the PDIs is to establish the threshold for compliance with the environmental flow regime defined in the RBMPs and to limit the occurrence of situations of temporary deterioration of water bodies quality only to prolonged drought natural phenomena situations (and not to scarcity situations), as set out in the WFD [6,28]. According to Article 18 of the Spanish Hydrological Planning Regulation [29], a prolonged drought situation allows the justified reduction of the environmental flows of water bodies established in the RBMPs.

A scarcity situation is defined as a temporal problem of lack of resources to meet the water demands associated with the different socioeconomic uses of water. Thus, the Scarcity Indicator (SI) is based on the relationship between the availability of resources and water demands, identifying the inability of the resources to meet the demands. Consequently, it serves as an instrument of assistance in decision making related to the management of water resources. The SI is an operational indicator, aiming at the progressive triggering of measures in order to postpone or avoid the occurrence of the most severe stages of scarcity, mitigating their adverse impacts to the several water uses [6,28].

The general methodology used for the definition of each of the two indicators is schematized in Figure 3.

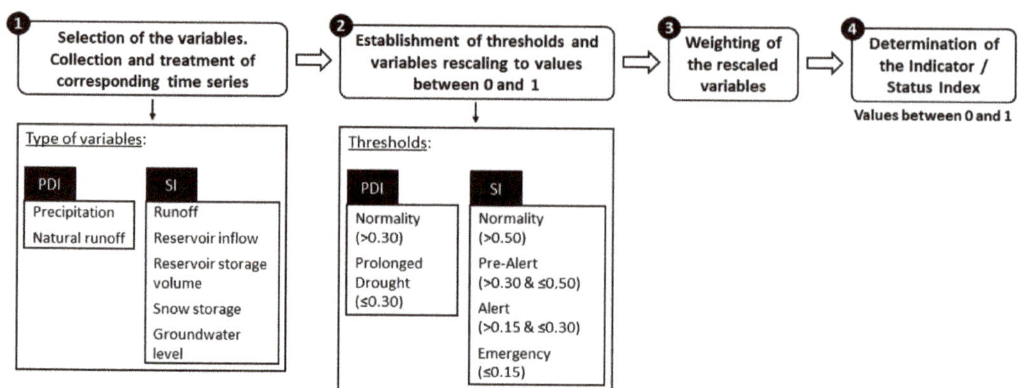

Figure 3. General methodology for the definition of both indicators (PDI and SI) (adapted from [20]).

The procedure begins with the selection of the hydrometeorological variables for each defined territorial unit of analysis. For the PDI, two types of variables may be considered: precipitation and runoff in natural regime. For the SI, the variables may be of various types, including runoff, reservoir inflow, reservoir storage volume, snow storage, and groundwater levels. These variables can be represented by their own values or by the corresponding standardized indices (e.g., SPI).

After the selection of the variables, they are rescaled into dimensionless variables varying between 0 and 1. In order for the rescaling of the variables to be performed, both indicators require a reference time series to be considered. The period between October 1980 and September 2012 was considered for all the reference series displayed in the 2018 PES-MS [20]. This 32-year hydrological monthly time series served as a sample to perform a statistical analysis of the data records, identifying the characteristics of the humid, normal and dry periods. Therefore, it allowed us to establish the predominant values/characteristics in the TMUs and enabled the analysis and characterization of later events by comparing it with the values of the reference series.

For both the Prolonged Drought and Scarcity indicators, the rescaling step requires the establishment of monthly thresholds for each of the 12 months of the year. One threshold was defined for PDI, corresponding the occurrence of prolonged drought, and three thresholds were defined for SI, corresponding to different limits of water scarcity stages (pre-alert, alert and emergency).

Subsequently, the rescaled variables were aggregated in a weighted way, producing a single indicator (also entitled as a Status Index, or "Indice de Estado" in Spain). This has a great advantage relative to the use of other type of indicators, namely simple standard indexes, because it enables to compare the indicator's results among different basins, despite their diverse geographical, climatic, water demands and other specific characteristics [6].

As referred before (in Section 1), the territorial units most suitable for the analysis and management of the two situations (prolonged drought and water scarcity) may be different. In the case of the 2018 PES-MS, the territorial units of analysis used for both indicators' definitions were those presented in Figure 2. Similarly, for the work developed and presented here, it was considered that the territorial units used for the Prolonged Drought and Scarcity indicator definitions in the Portuguese parts of the Minho and Lima river basins corresponded to their full country areas (also pictured in Figure 2).

In the Portuguese parts of the Minho and Lima river basins, the PDI and SI were defined for the period from October 1980 to September 2017, taking as reference time series the period from October 1980 to September 2012, such as in the 2018 PES-MS [20]. For the Lima river basin, the SI was defined for the period from October 1993 to September 2017, taking as reference time series the period from October 1993 to September 2012, as will be reported below (in Section 2.2.2).

The indicators obtained to the Portuguese parts of the Minho and Lima river basins were then compared, for each basin, with those obtained in the corresponding Spanish parts of the basins. In this context, the Portuguese parts of the Lima basin were compared with the ones defined to the Limia TMU. For the Minho basin, a similar comparison was made with the Miño-Bajo TMU, as this was the nearest territorial unit adjacent to the Portuguese part (see Figure 2).

The specific methodology used to define the Prolonged Drought and the Scarcity indicators, respectively, in the Portuguese parts of the Minho and Lima river basins are described below (in Sections 2.2.1 and 2.2.2), in more detail.

2.2.1. Prolonged Drought

In the Portuguese part of the Minho and Lima river basins, as considered in the Spanish part of the basins, two variables were considered in the definition of the PDI: (i) average monthly precipitation in the whole basin; and (ii) natural monthly runoff in one hydrometric station of the river basin hydrographic network.

Precipitation and runoff data recorded at stations of the Portuguese National Water Resources Information System (SNIRH in Portuguese acronym) monitoring network were used.

The monthly average precipitation in each Portuguese basin region was calculated through the Thiessen method, using the available data of the existing meteorological stations.

The hydrometric station chosen to represent the natural monthly runoff in the Portuguese part of the Lima river basin was "Pontilhão de Celeiros", located on Vez river (tributary at the right bank of the Lima river), its section having a sub-basin area of about 170 km^2. For the Minho hydrographic basin, the selected hydrometric station was "Segude", which corresponds to a sub-basin with an area of nearly 130 km^2 located on the Mouro river (tributary at the left bank of the Minho river) (Figure 4). To fill the missing data in the hydrometric stations, hydrological modeling using HEC-HMS software in a continuous simulation mode based on daily precipitation was performed.

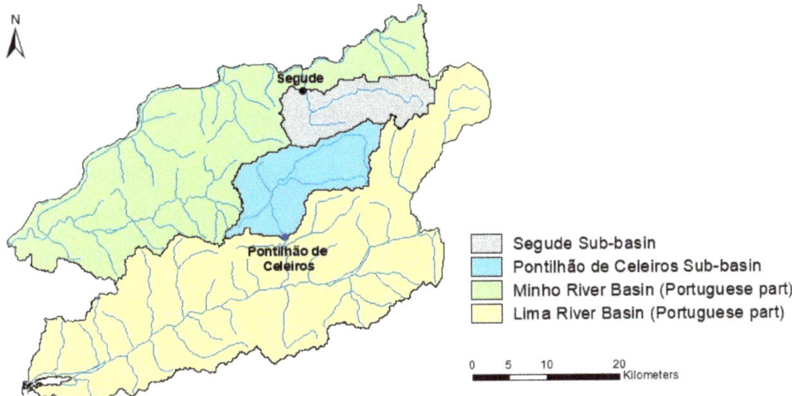

Figure 4. Relative location of the sub-basins at the sections of the hydrometric stations of "Pontilhão de Celeiros" (in the Lima river basin) and "Segude" (in the Minho river basin).

The two required variables (precipitation and natural runoff) were computed through the accumulated monthly records from the previous 12 months. Based on these, the two variables were transformed and translated into standardized values, namely, the Standardized Precipitation Index (SPI_{12}) for the variable precipitation, and the Standardized Runoff Index (SRI_{12}) for the variable natural runoff. For this purpose, the SPI and SRI were calculated by adjusting the precipitation and runoff data, respectively, to a gamma distribution [30].

After the selection and treatment of the variables, the SPI and SRI values were rescaled into dimensionless values ranging from 0 to 1. This rescaling process was achieved based on the following monthly reference values for each of the 12 months of the year:

- 1.0: Maximum value of SPI/SRI in the reference series (October 1980–September 2012);
- 0.5: Median value of SPI/SRI in the reference series (October 1980–September 2012);
- 0.3: Value of SPI/SRI equal to −1.2813 (cumulative probability of occurrence of 10%): value established as the prolonged drought threshold;
- 0.0: Minimum value in the reference series (October 1980–September 2012).

After rescaling, the two variables were aggregated, in a weighted way (considering 60% for the SPI and 40% for the SRI), resulting in the PDI. The referred threshold and the weighing values were the same used by Spain in the 2018 PES-MS [20], defined considering the comparison with the historical drought events in the Spanish part of the two basins. In accordance, values below 0.3 indicate the existence of a prolonged drought situation, whereas values above 0.3 indicate a situation of normality.

2.2.2. Scarcity

As for the PDI, the first step was the selection and treatment of the variable, or sets of variables, in each basin, that were considered the most representative of the availability of the water resources required to satisfy the different water demands. This selection was made considering the characterization and location of the most significant water demands and also the location of the water abstraction sources to meet those demands, namely, those related to urban water supply, agriculture, industry and tourism. In the Spanish part, the main water demands considered were relative to urban water supply and agriculture. In the Portuguese part of the Lima and Minho river basins, all demands (including agricultural) were not relevant compared to urban water supply, thereby only urban water demands were considered. In the two basins, although water is captured from surface and groundwater bodies, the main sources are surface water bodies.

In the Portuguese part of the Lima river, inflow to the Touvedo dam, located on the Lima river (Figure 5) was chosen as the variable representative of water availability. The Touvedo reservoir serves as the water abstraction source for the urban supply of a large cluster, including the main counties of the Lima river basin, namely Arcos de Valdevez, Ponte da Barca, Ponte de Lima, Viana do Castelo, as well for two counties of the Minho river basin, Caminha and Vila Nova de Cerveira.

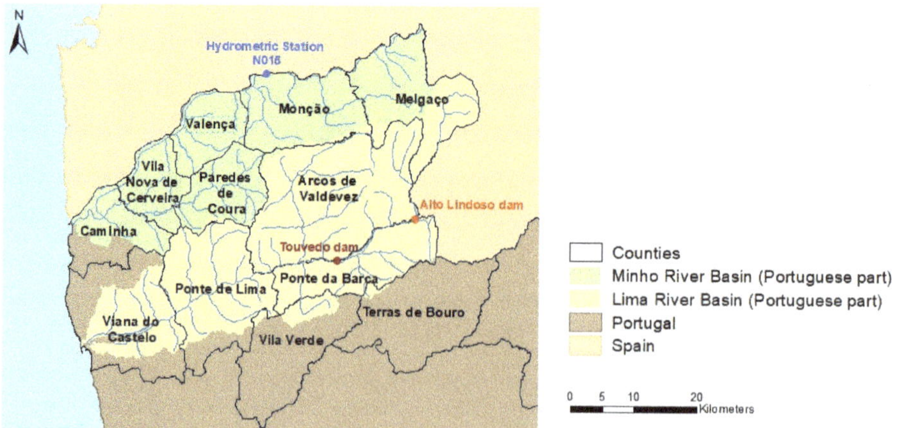

Figure 5. Counties of the Minho and Lima river basins and locations of the Touvedo dam (Lima) and hydrometric station N015—Rio Miño en Salvaterra do Miño (Minho).

As the Touvedo reservoir was only built in 1993, for the Lima river basin the SI analysis was carried out for the period from October 1993 to September 2017.

In the Minho river basin, the most important water abstraction source to urban supply is the Troporiz abstraction, which supplies Melgaço, Monção and Valença counties. Therefore, the variable selected to assess scarcity situations in the Portuguese part was the Spanish hydrometric station "N015–Rio Miño en Salvaterra do Miño", located in the Minho river (Figure 5), just 5 km upstream of the Troporiz abstraction.

In addition to the urban water supply demands, environmental flows associated to the Touvedo dam and to the Minho river abstraction section were also considered. Regarding the Minho river, the stretch where Spanish hydrometric station N015 is located was found to have two different environmental flow regimes definitions once compared of the two countries' definitions.

In fact, Portugal and Spain have adopted two different approaches in the development of the RBMPs, namely for the shared river basins parts: Spain defined environmental flows for all river stretches, while Portugal defined environmental flow regimes only for "heavily modified" water bodies located downstream of reservoirs [27,31–33]. Each country also uses different methods to define environmental flows, which, in addition to the different approaches mentioned above, could lead (namely as referred for Minho river) to a singular transboundary river body having two different environmental flow regime definitions.

In this study, the environmental flow regime considered was the one defined in the Spanish RBMP for the Minho river stretch where the water abstraction is located, since it is larger and thus consequently more challenging to satisfy in situations of scarcity.

The different scarcity situations considered were: Normality (absence of scarcity), Pre-alert (moderate scarcity), Alert (severe scarcity) and Emergency (serious scarcity). The corresponding monthly thresholds were defined for each of the months of the year based on the urban water supply demands considered for both territorial units and also considering the corresponding environmental flow regime, as follows:

- Pre-alert: median value of the reference series inflows per month;
- Alert: minimum inflow required to satisfy water demands and the environmental flows regime;
- Emergency: minimum inflow to satisfy water demands and 50% of the environmental flow regime.

Once the scarcity thresholds were defined, the variables were rescaled into dimensionless values between 0 and 1, based on the following reference values:

- 1: maximum value of the reference series inflows per month;
- 0.5: Pre-Alert threshold;
- 0.3: Alert threshold;
- 0.15: Emergency threshold;
- 0: minimum value of the reference series inflows per month. If the minimum value is greater than any one of the Emergency thresholds, the value 0 will correspond to the minimum Emergency threshold.

Since, in this case, only one variable was chosen to characterize scarcity situations (i.e., weighting factor equal to one) for each of the two basins, the rescaling of the variable resulted in the definition of the SI, to which the different situation range limits corresponded: values between 1 and 0.5 represented a situation of normality; values between 0.5 and 0.3 corresponded to a Pre-Alert situation; values between 0.3 and 0.15 represented an Alert situation; and values below 0.15 corresponded to an Emergency situation.

3. Results and Discussion

3.1. Prolonged Drought

By applying the methodology presented in Section 2.2.1, the PDI was calculated for the Portuguese part of the Minho and Lima river basins for the period from October 1980 to September 2017. The corresponding time evolution is depicted in Figure 6. In this figure, for each basin (Lima and Minho, respectively in Figure 6a,b), in addition to the evolution of the PDI in the Portuguese part, the indicator for the Spanish part of the basins is also presented.

(a)

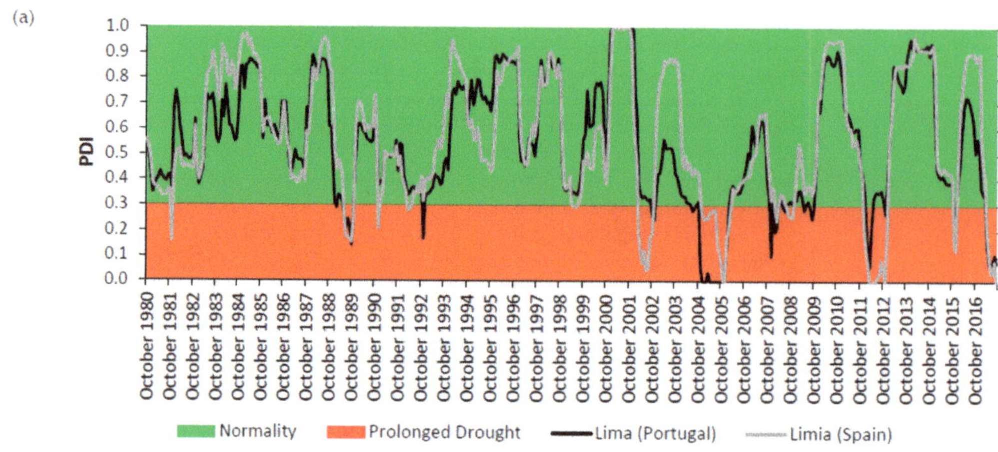

(b)

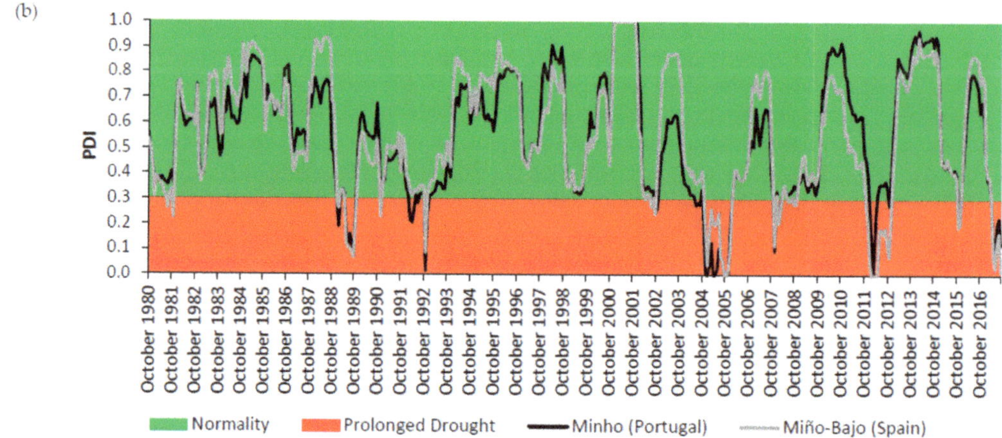

Figure 6. Prolonged Drought Indicator (PDI) for the period October 1980–September 2017 in (**a**) the Portuguese part (Lima) and the Spanish part (Limia) of the Lima river basin; and (**b**) in the Portuguese part (Minho) and the Spanish part (Miño-Bajo) of the Minho river basin.

Table 1 highlights the periods of months in which the situation of prolonged drought was signaled in each part of the two basins, as well as the total number of months of prolonged drought between October 1980 and September 2017.

From Table 1, it immediately follows that more months signaling prolonged drought occurred in the Spanish parts of the Minho and Lima river basins than in the respective Portuguese parts.

In general terms, it can be said that, except for the lengthy period of 16 months elapsed between 2004 and 2006, the Portuguese parts of the Minho and Lima river basins were marked by the occurrence of relatively short-term drought events (1 to 6 months), whose frequency of occurrence increased in recent years.

Table 1. Prolonged drought events in the Portuguese and Spanish parts of the Lima and Minho river basins (period October 1980 to September 2017).

	Lima (Portugal)	Limia (Spain)	Minho (Portugal)	Miño-Bajo (Spain)
PD events	February 1989 June 1989–November 1989 November 1992 October 2002–November 2002 July 2004–August 2004 November 2004–February 2006 December 2007 February 2008–March 2008 June 2008–July 2008 May 2009 August 2009–October 2009 January 2012–April 2012 November 2012 November 2015–December 2015 April 2017–September 2017	November 1981 June 1989–November 1989 December 1990 March 1992–May 1992 May 1999–July 1999 March 2002–October 2002 December 2004–March 2006 February 2008–March 2008 August 2008–November 2008 December 2011–December 2012 November 2015–December 2015 April 2017–September 2017	January 1989–February 1989 June 1989–November 1989 March 1992–May 1992 July 1992 November 1992–December 1992 October 2002–November 2002 July 2004–September 2004 November 2004–January 2006 February 2008–March 2008 January 2012–April 2012 November 2012 November 2015–December 2015 April 2017–September 2017	August 1981 November 1981 January 1989–February 1989 June 1989–November 1989 December 1990 November 1992 May 2002 August 2002 October 2002 December 2004–January 2006 December 2007 February 2008–March 2008 July 2008–August 2008 October 2008–November 2008 November 2011–December 2012 November 2015–December 2015 April 2017–September 2017
Total months in PD	50	65	50	58

In the Portuguese parts of the Minho and Lima river basins, the PDI values (Figure 6) show that the most severe drought episode in intensity and duration occurred between late 2004 and the beginning of 2006.

In the Spanish part of both basins, two large prolonged drought events were signaled, namely the one that occurred between the end of 2004 and the beginning of 2006 and the one that occurred between the end of 2011 and the end of 2012. In the Lima river basin (Limia), the first (2004–2006) episode was the longest, lasting 16 months. However, the later (2011–2012) event was the one with the greatest intensity. In the Minho river basin, the two events presented similar intensities and durations.

In general, despite differences in intensity and duration, there is correspondence between the periods signaled as suffering prolonged drought in the Portuguese and Spanish parts of both basins between 1980/1981 and 2016/2017.

By the observation of Figure 6, it can also be inferred that the major drought events, namely those occurring in 2004/2005–2005/2006, 2011/2012 and 2016/2017, affected both international basins. It should be noted that, in the most recent occurrence, the duration and intensity of the drought events were similar in both parts of the Minho and Lima river basins.

3.2. Scarcity

Figure 7 presents the evolution of the SI for the Lima and Minho River Basins, displaying both the Portuguese and the Spanish values, respectively, obtained in this study and reported in 2018 PES-MS [20]. As previously mentioned, for the Portuguese Lima river basin part, the period of analysis was from October 1993 to September 2017. For this reason, although the indicator was computed for the period from October 1980 to September 2017 for the Spanish part of the basin, only the period from October 1993 onward was presented.

Considering the period of analysis for each basin, Table 2 presents, for each part of the basins, the number of months of prevalence corresponding to each scarcity stage, as well as the respective percentages, considering the entirety of the respective period of analysis.

Through Figure 7 and Table 2, it is possible to observe that, when compared with the Portuguese part, the Spanish part of the Lima river basin had a greater exposure to scarcity situations. For the period considered (October 1993–September 2017), the Portuguese part only exhibited situations of Normality and Pre-Alert. In contrast, in the Spanish part of the basin, there was frequent occurrence of the Alert and Emergency statuses, particularly in the last decade of the period (2008–2017).

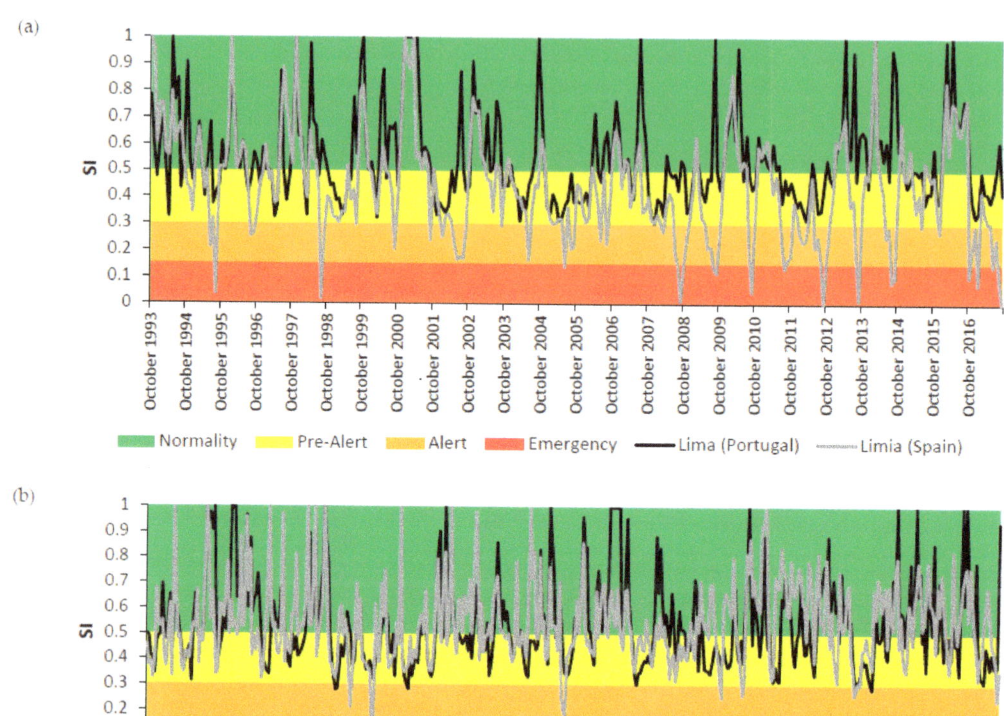

Figure 7. Scarcity Indicator (SI) in (**a**) the Portuguese part (Lima) and the Spanish part (Limia) of the Lima river basin for the period October 1993–September 2017; and in (**b**) the Portuguese part (Minho) and the Spanish part (Miño-Bajo) of the Minho river basin for the period October 1980–September 2017.

Table 2. Distribution of the scarcity status in the Portuguese and Spanish parts of the Lima and Minho river basin.

	Lima (Portugal)		Limia (Spain)		Minho (Portugal)		Miño-Bajo (Spain)	
Status	**Months**	**(%)**	**Months**	**(%)**	**Months**	**(%)**	**Months**	**(%)**
Normality	155	53.8	109	37.8	223	50.2	248	55.9
Pre-Alert	133	46.2	116	40.3	216	48.7	188	42.3
Alert	0	0.0	45	15.6	5	1.1	7	1.6
Emergency	0	0.0	18	6.3	0	0.0	1	0.2

With regard to the Minho river basin, both parts had low exposure to scarcity situations, presenting very few cases falling under the scope of the Alert status, between October 1980 and September 2017. It should also be noted that the Spanish part, throughout the analysis series, presented only one month in Emergency status.

4. Conclusions

To minimize water conflicts and prevent drought- and scarcity-related impacts on international river basins, it is essential that both countries have similar planning and management approaches. In this regard, Portugal and Spain are not yet in similar stages. Notably, the definition of comparable drought indicators to be applied in common by the two countries in the shared river basins is missing.

In this context, the goal of the present work was the definition of common drought indicators in the shared river basins as a tool for the better coordination of drought and water scarcity management in the Iberian Peninsula. For this purpose, examining the Minho and Lima river basins as case study, this work presented the definitions of the indicators in the Portuguese part of the Minho and Lima river basins similar to those defined in the Spanish part of the two basins, namely the Prolonged Drought and Scarcity indicators.

The resulting indicators showed that, in terms of prolonged drought and despite different intensities and durations, there was a general coincidence between the periods of prolonged drought in the Portuguese and Spanish parts of the Lima and Minho river basins between 1980/1981 and 2016/2017, with the major drought events being signaled in the two parts of each basin.

Concerning scarcity situations, both the Portuguese and Spanish parts of the Minho river basins and the Portuguese part of the Lima river basin have low exposure to scarcity situations. The only exception is the Spanish part of the Lima river basin, which has a high level of exposure to this situation.

Based on the results presented, namely due to the apparent adequacy of the indicators obtained for Spanish and Portuguese parts of the Minho and Lima river basins, it can be concluded that the general Spanish methodology presented may be considered to be a valid starting point for the Portuguese part of Minho and Lima river basins and that it has also the potential to be applied in the others Portuguese basins.

The advantage of using these indicators instead of the use of more conventional singular indicators, such as the SPI or other hydrological indicators, corresponds to the fact of diagnosis and indicator results being more clearly comparable between the different basins, despite their proper characteristics. Moreover, the definition of similar indicators has extreme importance for transboundary drought and scarcity management since, common definitions enable the common characterization and comparison of droughts and scarcity events in the Portuguese and Spanish parts of the shared river basins. That may allow the common planning, management and monitoring of droughts and scarcity situations between the two countries.

In addition to a better integration of both countries with WFD principles and goals, the drought and scarcity common indicators' definitions and management shall lead to the improvement of the Albufeira Convention implementation by contributing not only to the envisaged redefinition of the MFR but also to an agreement on the definition of environmental flows for common stretches of the transboundary rivers.

Nevertheless, it should be emphasized that some work remains to be developed to improve the current system of indicators to be used in common by the two countries. In fact, despite being in a way quite integrated with the WFD (distinguishing prolonged drought situations from scarcity situations), the use of the Spanish system of indicators has some recognized limitations, cautions and reserves that must be considered [34], namely, concerning the PDI. In the Spanish system, the definition of the PDI only considers the intensity and not the duration of the events; the arbitrary definition of the threshold for prolonged drought is equal to an indicator value of 0.30. This definition should be agreed with Portugal, because the measures envisaged to be applied by Spain for prolonged drought—the call in Article 4.(6) of the WFD and reduction of environmental flows—may not only affect Portugal but also in fact themselves constitute the consequences of prolonged drought events. Therefore, the management of prolonged drought situations should be performed in order to achieve a maximum delay on the use of Article 4.(6) and on the reduction of the environmental flows, which ought not to be exacerbated.

Author Contributions: Conceptualization, R.M., M.C. and J.M.; methodology development and application, M.C.; result analysis and validation, R.M., M.C. and J.M.; writing—original draft preparation, M.C. and J.M.; writing—review and editing, R.M., M.C. and J.M.; supervision, R.M. All authors have read and agreed to the published version of the manuscript.

Funding: This research was supported by the RISC_ML project (reference: 0034_RISC_ML_6_E), co-funded by the European Regional Development Fund (ERDF) through the Interreg V-A Spain-Portugal Program (POCTEP) 2014–2020.

Institutional Review Board Statement: Not applicable.

Informed Consent Statement: Not applicable.

Data Availability Statement: The data used in this study are available in the publicly accessible repositories: SNIRH database (https://snirh.apambiente.pt/, accessed on 20 October 2021) and SAIH Miño Sil (http://saih.chminosil.es/, accessed on 20 October 2021).

Conflicts of Interest: The authors declare no conflict of interest.

References

1. Van Lanen, H.A.; Vogt, J.V.; Andreu, J.; Carrão, H.; De Stefano, L.; Dutra, E.; Feyen, L.; Forzieri, G.; Hayes, M.; Inglesias, A.; et al. Climatological risk: Droughts. In *Science for Disaster Risk Management 2017: Knowing Better and Losing Less*, 1st ed.; Poljansek, K., Marin Ferrer, M., De Groeve, T., Clark, I., Eds.; Publications Office of the European Union: Luxembourg, 2017; ISBN 978-92-79-60679-3.
2. Kalis, G. Droughts. *Annu. Rev. Environ. Resour.* **2008**, *33*, 85–118. [CrossRef]
3. Van Loon, A.F.; Van Lanen, H.A. Making the distinction between water scarcity and Drought using an observation-modeling framework. *Water Resour. Res.* **2013**, *49*, 1483–1502. [CrossRef]
4. Vivas, E. Assessment and Management of Drought and Scarcity Situations. Application to the Guadiana Case. Ph.D. Thesis, Faculty of Engineering of the University of Porto, Porto, Portugal, 2011. (In Portuguese).
5. Wilhite, D.A.; Sivakumar, M.V.; Pulwarty, R. Managing drought risk in a changing climate: The role of national drought policy. *Weather. Clim. Extrem.* **2014**, *3*, 4–13. [CrossRef]
6. Hervás-Gámez, C.; Delgado-Ramos, F. Drought Management Planning Policy: From Europe to Spain. *Sustainability* **2019**, *11*, 1862. [CrossRef]
7. European Commission. Directive 2000/60/EC of the European Parliament and of the Council of 23 October 2000 establishing a framework for Community action in the field of water policy. L 327 2000, 01-0073. In *Official Journal of the European Communities, L327/1*; European Parliament and the Council of the European Union: Brussels, Belgium, 2000. Available online: https://eur-lex.europa.eu/legal-content/EN/TXT/?uri=CELEX:32000L0060 (accessed on 20 October 2021).
8. Zikos, D.; Hagedorn, K. Competition for Water Resources from the European Perspective. In *Competition for Water Resources: Experiences and Management Approaches in the US and Europe*, 1st ed.; Ziolkowska, J., Peterson, J., Eds.; Elsevier: Cambridge, MA, USA, 2017; pp. 19–35. [CrossRef]
9. European Commission. *Communication from the Commission to the European Parliament and the Council—Addressing the Challenge of Water Scarcity and Droughts in the European Union. COM(2007) 414 Final*; Commission of the European Communities: Brussels, Belgium, 2007. Available online: https://eur-lex.europa.eu/legal-content/EN/ALL/?uri=CELEX%3A52007DC0414 (accessed on 20 October 2021).
10. European Commission. *Communication from the Commission to the European Parliament, the Council, the European Economic and Social Committee and the Committee of the Regions—Report on the Review of the European Water Scarcity and Droughts Policy. COM(2012) 672 Final*; European Commission: Brussels, Belgium, 2012. Available online: https://eur-lex.europa.eu/legal-content/EN/TXT/?uri=CELEX:52012DC0672 (accessed on 20 October 2021).
11. European Commission. *Communication from the Commission to the European Parliament, the Council, the European Economic and Social Committee and the Committee of the Regions—Forging a Climate-Resilient Europe—the New EU Strategy on Adaptation to Climate Change. COM(2021) 82 Final*; European Commission: Brussels, Belgium, 2021. Available online: https://eur-lex.europa.eu/legal-content/EN/TXT/?uri=COM:2021:82:FIN (accessed on 20 October 2021).
12. European Commission. *Report from the Commission to the European Parliament and the Council—Third Follow up Report to the Communication on Water Scarcity and Droughts in the European Union COM (2007) 414 Final. COM(2011) 133 Final*; European Commission: Brussels, Belgium, 2011. Available online: https://eur-lex.europa.eu/legal-content/EN/TXT/?uri=CELEX:52011DC0133 (accessed on 20 November 2021).

13. Maia, R.; Vicente-Serrano, S. Drought Planning and Management in the Iberian Peninsula. In *Drought and Water Crises: Integrating Science, Management, and Policy*, 2nd ed.; Wilhite, D., Pulwarty, R., Eds.; CRC Press: Boca Raton, FL, USA, 2017; pp. 481–506. [CrossRef]
14. DR. Resolution of the Assembly of the Republic n° 66/99—Convention on Cooperation for the Protection and Sustainable Use of the Waters of the Portuguese-Spanish River Basins and the Additional Protocol. In *Diário da República n° 191/1999, Série I-A de 1999-08-17 (Portuguese Government Official Journal)*; Assembleia da República: Lisboa, Portugal, 1999. Available online: https://dre.pt/dre/detalhe/resolucao-assembleia-republica/66-1999-434660 (accessed on 20 November 2021). (In Portuguese)
15. DR. Resolution of the Assembly of the Republic no. 62/2008—Protocol of Review of the Convention on Cooperation for the Protection and Sustainable Use of the Waters of the Portuguese-Spanish River Basins (Albufeira Convention) and the Additional Protocol. In *Diário da República n° 222/2008, Série I de 2008-11-14 (Portuguese Government Official Journal)*; Assembleia da República: Lisboa, Portugal, 2008; pp. 7961–7969. Available online: https://dre.pt/dre/detalhe/resolucao-assembleia-republica/62-2008-439852 (accessed on 20 November 2021). (In Portuguese)
16. Chainho, P.; Martínez-Fernández, J. Iberian waters under European protection: Soft water in hard stone? In Proceedings of the X Congresso Ibérico de Gestão e Planeamento da Água, Coimbra, Portugal, 6–8 September 2018; ISBN 978-84-944788-5-7. (In Portuguese and Spanish).
17. Arca, I. The Albufeira Convention: An analysis on the occasion of the Twentieth Anniversary of its Adoption. *Rev. Catalana De Dret Ambient.* **2019**, *10*, 1–38. (In Spanish) [CrossRef]
18. Estrela, T.; Sancho, T. Drought management policies in Spain and the European Union: From traditional emergency actions to Drought Management Plans. *Water Policy* **2016**, *18*, 153–176. [CrossRef]
19. Spanish Government. Order TEC/1399/2018, of November 28, approving the review of the special drought plans. In *BOE núm. 311, de 26 de Diciembre de 2018 (Spanish Government Official Journal)*; Ministerio para la Transición Ecológica: Madrid, Spain, 2018. Available online: https://www.boe.es/diario_boe/txt.php?id=BOE-A-2018-17752 (accessed on 20 November 2021). (In Spanish)
20. Confederación Hidrográfica del Miño-Sil. *Special Plan of Action in Situations of Alert and Eventual Drought of the Spanish Part of the Miño-Sil River Basin District: Memory*; CHMiño-Sil: Ourense, Spain, 2018. Available online: https://www.chminosil.es/es/chms/planificacionhidrologica/nuevo-plan-especial-de-sequia/plan-especial-en-situaciones-de-alerta-y-eventual-sequia-de-la-demarcacion-hidrografica-del-mino-sil-pes-version-definitiva-aprobada-y-vigente (accessed on 20 November 2021). (In Spanish)
21. Comissão Permanente de Prevenção, Monitorização e Acompanhamento dos Efeitos de Seca. Prevention, Monitoring and Contingency Plan for drought Situations. 2017. Available online: https://apambiente.pt/agua/plano-de-prevencao-monitorizacao-e-contingencia-para-situacoes-de-seca (accessed on 20 November 2021). (In Portuguese).
22. Maia, R. The Guadiana Case Study. In *GAR Special Report on Drought 2021*; United Nations Office for Disaster Risk Reduction: Geneva, Italy, 2021. Available online: https://www.undrr.org/publication/guadiana-case (accessed on 20 December 2021).
23. Expresso. The Water No Longer Flows in the River: Images of Desolation Caused by Drought and Spain. Available online: https://expresso.pt/sociedade/2019-10-12-A-agua-ja-nao-corre-no-rio-as-imagens-da-desolacao-provocada-pela-seca-e-por-Espanha (accessed on 20 October 2021). (In Portuguese)
24. Ministério do Ambiente e da Ação Climática. Flows of the Tejo River. Available online: https://www.portugal.gov.pt/pt/gc22/comunicacao/comunicado?i=caudais-do-rio-tejo (accessed on 20 December 2021). (In Portuguese)
25. Do Ó, A. Transboundary Drought Management in the Guadiana: Applying the Conflict Risk Index. In Proceedings of the 5° Congresso Internacional de Ordenamento do Território—Gestão Partilhada de Recursos Hídricos Internacionais, Lisboa, Portugal, 27–28 October 2010.
26. RISC_ML: Prevention of Flood and Drought Risks in the Minho-Lima International River Basins. Available online: http://risc-ml.eu (accessed on 20 December 2021).
27. Agência Portuguesa do Ambiente. *Minho and Lima River Basin Management Plan (RBD1) 2016–2021. Part 2—Characterization and Diagnostics (Annexes)*; APA: Lisboa, Portugal, 2016. Available online: https://apambiente.pt/agua/2o-ciclo-de-planeamento-2016-2021 (accessed on 20 November 2021). (In Portuguese)
28. Ministério para la Transición Ecológica. *Technical Instruction for the Preparation of Special Drought Plans and the Definition of the Global System of Indicators of Prolonged Drought and Scarcity*; MITECO: Madrid, Spain, 2017. Available online: https://www.miteco.gob.es/es/agua/participacion-publica/PP-Agua-Orden-instruccion-tecnica-elaboracion-planes-especiales-sequia.aspx (accessed on 20 November 2021). (In Spanish)
29. Spanish Government. Real Decreto 907/2007—Hydrological Planning Regulation. In *BOE Núm. 162, de 7 de Julio de 2007 (Spanish Government Official Journal)*; Ministerio de Medio Ambiente: Madrid, Spain, 2007. Available online: https://www.boe.es/buscar/doc.php?id=BOE-A-2007-13182 (accessed on 20 November 2021). (In Spanish)
30. McKee, T.; Doesken, N.; Kleist, J. The relationship of drought frequency and duration to time scales. In Proceedings of the Eight Conference on Applied Climatology, Anaheim, CA, USA, 17–22 January 1993.
31. Agência Portuguesa do Ambiente. *Douro River Basin Management Plan (RBD3) 2016–2021. Part 2—Characterization and Diagnostics (Annexes)*; APA: Lisboa, Portugal, 2016. Available online: https://apambiente.pt/sites/default/files/_SNIAMB_Agua/DRH/PlaneamentoOrdenamento/PGRH/2016-2021/PTRH3/PGRH_2_RH3_Parte2_Anexos.pdf (accessed on 20 November 2021). (In Portuguese)

32. Agência Portuguesa do Ambiente. *Tejo and Ribeiras do Oeste River Basin Management Plan (RBD5) 2016–2021. Part 2—Characterization and Diagnostics (Annexes)*; APA: Lisboa, Portugal, 2016. Available online: https://apambiente.pt/sites/default/files/_SNIAMB_Agua/DRH/PlaneamentoOrdenamento/PGRH/2016-2021/PTRH5A/PGRH_2_RH5A_Parte2_Anexos.pdf (accessed on 20 November 2021). (In Portuguese)
33. Agência Portuguesa do Ambiente. *Guadiana River Basin Management Plan (RBD7) 2016–2021. Part 2—Characterization and Diagnostics (Annexes)*; APA: Lisboa, Portugal, 2016. Available online: https://apambiente.pt/sites/default/files/_SNIAMB_Agua/DRH/PlaneamentoOrdenamento/PGRH/2016-2021/PTRH7/PGRH_2_RH7_Parte2_Anexos.pdf (accessed on 20 November 2021). (In Portuguese)
34. Fundación Nueva Cultura del Agua (FNCA). *FNCA Observations to the "Technical Instruction for the Preparation of Special Drought Plans and the Definition of the Global System of Indicators of Prolonged Drought and Scarcity"*; FNCA: Zaragoza, Spain, 2018. (In Spanish)

Article

Adaptation: A Vital Priority for Sustainable Water Resources Management

Elpida Kolokytha

Department of Civil Engineering, School of Engineering, Aristotle University of Thessaloniki, 54124 Thessaloniki, Greece; lpcol@civil.auth.gr

Abstract: Sustainability in terms of water management implies the study of all interrelated parameters (social, environmental, economic, engineering and political) in a comprehensive way. Although Greece is presented in the international rankings as a water-rich country, it has significant water problems due to its high temporal and spatial distribution of water resources and its unsustainable management practices characterized by a fragmented and sector-oriented water management system. This problem has been significantly improved by the adoption of the EU WFD and the development of management plans at the river basin scale. Nevertheless, because of the climate change effects, there is still a long way to go, and radical changes are needed in order to reach sustainability. Adaptation is a vital response toward sustainability. The Mygdonia agricultural basin is a case study of a highly negative water balance system that highlights the shortcomings of both water management and adaptation in Greece. Analysis of the hydrology of the basin, as well as the climate projections until 2100, revealed the urgent need for concerted action. A set of different development adaptation strategies was applied and assessed concerning their effectiveness. According to the outputs of this research, integrated watershed management is a prerequisite for a successful adaptation policy. Radical reform is needed in the agricultural sector by decreasing the agricultural land and changing crops. Demand management is the solution rather than focusing on supply options.

Keywords: climate crisis; water adaptation; Greece; Koronia lake; sustainability; Mygdonia Basin

Citation: Kolokytha, E. Adaptation: A Vital Priority for Sustainable Water Resources Management. *Water* **2022**, *14*, 531. https://doi.org/10.3390/w14040531

Academic Editor: Maria Mimikou

Received: 28 December 2021
Accepted: 5 February 2022
Published: 11 February 2022

Publisher's Note: MDPI stays neutral with regard to jurisdictional claims in published maps and institutional affiliations.

Copyright: © 2022 by the author. Licensee MDPI, Basel, Switzerland. This article is an open access article distributed under the terms and conditions of the Creative Commons Attribution (CC BY) license (https://creativecommons.org/licenses/by/4.0/).

1. Introduction

The scientific question on whether the observed recent climate changes are anthropogenic or exclusively natural occurring over time as the natural cycle of climate change has been the subject of several studies in the last few decades [1–3].

The Intergovernmental Panel on Climate Change (IPCC) has recently released the Physical Science Basis report (August 2021), where it is documented that "climate change is already affecting nearly every part of the planet, and human activities are unequivocally the cause" [4]. It is imperative to immediately take action.

Regardless of the main reasons, what the whole scientific community agrees on is that climate change exists and that a new reality is here to be dealt with.

Water is the most vital component of life and is critical for almost all economic activities; as such, it is central to the achievement of sustainable development. The Global Risks Report of 2020 ranks environmental issues (among them, extreme weather events natural disasters, water crisis, failure regarding climate action) first on a list of the top global risks in terms of the impact on humanity [5]. According to the UN, climate change is projected to increase the number of water-stressed regions and exacerbate shortages in already water-stressed regions [6]. Alteration of the water cycle (quantity and quality) and an increase in extremes events are major impacts of climate change on freshwater resources. The planet will face a 40% shortage in water supply by 2030. Hydrological disasters, floods and storms accounted for 44% and 28%, respectively, of all disaster events from 2000 to 2019, affecting 1.6 billion people worldwide [5].

Climate change is characterized by great uncertainty, affecting the development model of a country directly and decisively, with significant differences in time and space at all levels, namely, local, regional, national and global. In order to reverse or halt the severe consequences of the current climate crisis, we need to work collectively and understand the depth and complexity of this crisis. Two types of responses for climate crisis mitigation and adaptation need to be applied concurrently. Mitigation [7] addresses the root cause of climate change (accumulation of greenhouse gases in the atmosphere), whereas adaptation addresses the impacts of climate change. Adaptation, in brief, anticipates the adverse effects of climate trends and takes appropriate action to prevent or minimize the damage they can cause [8–15]. Since mitigation reduces the rate, as well as the magnitude, of the root cause (warming), it also increases the time available for adaptation to a particular level of the climate crisis.

In the 21st century, "a new global theory" for water and its management is needed. Until recently, the hydrologic record of the past was the best guide for the future [16]. However, due to the increase in the rate of extreme events, as well as the non-stationarity and the great vulnerability and uncertainty in hydrological projections, we need to move from a solely technical and engineering management of water to a clear understanding of the complicated links between land, forest, agriculture, biodiversity, energy, health and true integration of the human dimension. We need to make water management more adaptive and flexible to be operational under fast-changing global socio-economic and climate-sensitive conditions [17,18]. A major issue in this effort is the reassessment of the global consumption and production model to manage food security, water scarcity and sustainable development through effective adaptations in agriculture.

In their quest for sustainable development, policymakers have to make trade-offs between the benefits and costs of adaptation measures, opinions on how much risk is socially acceptable and other development objectives [19,20].

The objective of the current research was to assess the climate impacts on water management in basins with severe water deficits by providing a better understanding of the adaptation options at a local level. More specifically, demand and supply adaptation strategies are explored in river basins with negative water balances and intense agricultural activity. The implementation of adaptation options was achieved by using a comprehensive analysis of both hydrologic and water management methods. This approach amplified the premises of sustainability, reflected new paradigms and practices and explained the opportunities for innovative approaches in water resources management. The case study of Mygdonia Basin was used as a representative example, as it is a highly water-stressed agricultural basin with an already negative water regime. Similar cases are encountered, both in other basins in Greece and in the Mediterranean. This article can act as a useful decision-making tool for policymakers to implement adaptation solutions to manage water resources, taking into account climate impacts in the area under study in a sustainable way.

2. Materials and Methods

2.1. Water Resources and Uses in Greece

Greece is located in the eastern part of the Mediterranean Basin, with the most extensive coastline in Europe of 15,000 km along the Mediterranean Sea. The climate is characterized by mild and rainy winters, relatively warm and dry summers and long sunshine duration almost all year long. Although the country is claimed to have adequate water reserves at 6471 m^3/capita/year [21], it suffers from a high temporal and spatial distribution of the water supply, which causes significant water shortages in specific regions in Greece [22,23]. The particular geomorphological conditions with the wet mountainous region concentrated along the backbone of the country, the rather dry long coastline and the numerous islands scattered in the Ionian and Aegean seas are responsible for the uneven distribution of the water supply. As a result, plentiful water can be found in the mountains flowing into the sea creating small torrents and rivers during winter, with almost no flow during summer in the dry period [24]. Moreover, there is high spatial and temporal water demand. Greece is an agricultural country, with 84% of its consumption belonging to the agricultural sector, with strong tourism in the islands and high seasonal water demand in the summer. High urbanization, with half of the population concentrated in two cities (Athens and Thessaloniki), where competition for water for economic activities is high, is another reason for the temporal and spatial water demand in Greece. Furthermore, the dependence of the country on the 16–20% of waters imported by four transboundary rivers reveals the extent of the water problem.

2.2. Demand Management versus Supply Management

Demand management has gone largely unaddressed in Greece since most water services focused on infrastructure development rather than on water conservation. This supply-oriented water policy for all these years was based on the "notion" that the solution to the water problem relies unilaterally on the country's capacity for engineering solutions to divert, construct and bring water to where it is needed, no matter how far and how costly this may be, has resulted in negative water balances and the depletion of groundwater reserves in many basins (Central Macedonia, Thessaly, Aegean Islands) [25]. Moreover, tools such as water pricing, especially in the agricultural sector, were only considered a viable option under the condition that all other supply options were exhausted, and overexploitation of surface and groundwater resources had resulted in water depletion. This hydrological reality, together with the fact that climate change will deteriorate the water reserves, calls for a drastic change in managing water for all uses. In such cases of drained water basins, the efficacy of measures that reduce/limit the use of water is questionable. It is highly likely that more drastic measures need to be taken, such as a change in the relevant economic activities and the suspension of the most water-consuming activities among them.

2.3. Water Climate Impact Projections in the Mediterranean and Greece

The Mediterranean Basin is a region that is already greatly affected by climate change [26], which is expected to remain among the "hotspot" regions most affected by climate change in the future, particularly when it comes to precipitation and the hydrological cycle [27–30]. The climate is changing in the Mediterranean Basin faster than global trends [26]. It is expected that heat waves will intensify in duration and peak temperatures, as well as heavy rainfall events, are likely to also intensify by 10–20% in all seasons except summer [31,32]. Precipitation and temperature changes are expected to increase crop water requirements [33] while putting food security in peril [34].

Despite strong regional variations, summer rainfall will likely be reduced by 10 to 30% in some regions, increasing existing water shortages and desertification and decreasing agricultural productivity [33,35]. As a typical Mediterranean country, Greece will experience these impacts. Water resources seem to be particularly affected by climate change in Greece, as it is reported that Greece ranked 26th among the countries that experienced severe water stress in 2019 and this water stress is highly likely to get worse by 2050 [36].

The Bank of Greece [37] published a detailed assessment of climate projections over Greece. In this report, in order to capture the possible changes in the water potential of the country until the year 2100, hydrological balance components were estimated for the periods of 2021–2050 and 2071–2100 using the emission scenarios A1B, A2 and B2. Details on the emission scenarios can be found in [6]. The results indicated significant changes in the hydrological components for each possible scenario. Specifically, the comparison regarding the changes in rainfall volume and total water potential (surface runoff and groundwater discharge) per climatic scenario in the whole Greek territory under current and future conditions predicted a reduction in rainfall ranging between 3% to 7% and a total water potential reduction (surface runoff and groundwater discharge) by 7–20% for the period of 2021–2050. Concerning the period of 2071–2100, the reduction will continue and most probably will be even higher, ranging from 14% to 22% regarding rainfall and between 30–54% regarding the water potential for the whole country [37]. These findings are in line with the IPCC 2014 report, which predicted (based on climate model A1B) that in the eastern part of the Mediterranean Basin (where Greece is located), rainfall in 2080–2099 will show a decrease of more than 20% compared with the period of 1980–1999. Furthermore, the annual precipitation is expected to decrease in most Mediterranean areas, including Greece, with the annual number of precipitation days being decreased. It is noted that the risk of summer drought is expected to rise, especially in southern Greece, while the duration of the snow season is very likely to be shorter [38,39].

In relation to temperature, according to the IPCC 2014 report, the annual mean temperatures, as well as the evapotranspiration in Greece, are likely to increase more than the global mean, especially regarding maximum summer temperatures.

In Greece, an increase in irrigation and tourist needs, as well as in the pollution load, is expected in the near future [37,40]. Vulnerability to climate change for the period 2050–2100 in comparison with the period 1961–1990 was also assessed. Greece shows a high vulnerability in Central Macedonia and the Western Peloponnese and moderate vulnerability in Thrace, Thessaly, Attica and Rhodes [41–44].

It is evident that the already disturbed water balances in the water basins in Greece will be further accelerated based on the future projections of climate impacts. Obviously, as precipitation has a strong local/regional component, this acceleration will have different results and will be shown first in the most vulnerable regions.

2.4. The Mygdonia Water Basin

Mygdonia Basin is located in Central Macedonia, 11.5 km northeast of the city of Thessaloniki. It occupies an area of 2061.48 km^2, with a population of approximately 65,000 inhabitants (ELSTAT2011), and hydrologically belongs to the Water District of Central Macedonia (GR10). It is surrounded by mountains with an altitude of 600–1200 m and the climate is a typical Mediterranean one. The surface flow is seasonal coming from distributed streams in winter, whereas during the dry season, their flow is reduced or almost non-existent.

Mygdonia Basin is a protected wetland according to the Ramsar Convention, with a complex water system comprising Lake Koronia (western part of the basin, 1278 km^2), Lake Volvi (eastern part of the basin 783.48 km^2) and the Mygdonia groundwater aquifer (Figure 1). The Ramsar Convention encourages the designation of sites containing representative, rare or unique wetlands or wetlands that are important for conserving biological diversity. The Koronia and Volvi wetlands support endemic fish, nesting waterbirds and large numbers of wintering birds, including Anatidae (geese, ducks, swans, etc.). Several nationally rare or endangered aquatic plants also occur here [45].

Figure 1. Map of the reference area provided by Malamatatis D. in his PhD.

The mild climate and the fertile soil favor irrigation of these lands and have contributed to the rapid development of agriculture. The economy of the area consists of small local enterprises serving the needs of the local communities. Besides agriculture, residents are mainly employed in livestock, while the secondary section of the economy mainly involves wood and construction enterprises. This information is given by the responsible Koronia–Volvi Management Body.

The overflow of Lake Koronia drains into Lake Volvi through the Derveni stream (Figure 1). The overflow of Lake Volvi drains into the Strymonikos Gulf through the Richios stream. The groundwater flows from the Koronia sub-catchment to the Volvi sub-catchment, and then a part of this discharge outflows to Lake Volvi and another one outflows to the Strymonikos Gulf. There is no flow interaction between the groundwater and Lake Koronia, as the bottom of the lake is impermeable [46].

It is a predominantly agricultural water basin in the Mediterranean region, with 95% of the basin's water being used for agricultural purposes [45], which suffers from unsustainable water management practices. During the last few decades, Lakes Volvi and Koronia, along with the Mygdonia Basin aquifer, have undergone severe quantitative and qualitative degradation due to past industrial, agricultural and urban activities. In particular, the water depth of Lake Koronia has progressively decreased since 1970, resulting in complete depletion in the summer of 2008. Lake Volvi, as a larger and deeper lake compared with Lake Koronia, experienced a smaller reduction of its depth. Moreover, the limited recharge to the Mygdonia Basin aquifer and the over-pumping for irrigation caused a significant drawdown of the groundwater table [47]. Central Macedonia (GR10) faces groundwater quantity pressure, as about 25% of the groundwater bodies are characterized to be in a poor/bad quantity state [44].

The environmental problems of the Mygdonia Basin were initially recognized in 1995 when an episode of mass fish deaths took place in Lake Koronia. The environmental collapse of Lake Koronia resulted in the drafting of the "Master Plan for the restoration of Lake Koronia" in 1998. The Master Plan had been oriented toward large-scale infrastructures and a water transfer scheme from the River Aliakmon, which flows in a neighboring catchment. Several of the proposed solutions in the Master Plan raised objections from both the central administration and international institutions since they were hard engineering projects that would cause considerable environmental impacts in the area, mainly altering the Ramsar protected ecosystem and the hydrodynamics of the water systems. In 2004, a "Revised Restoration Plan of Lake Koronia" was carried out to review the first Master Plan.

The overexploitation of the surface water system (Lakes Koronia and Volvi) and groundwater resources during the previous decades, along with the projected decrease in the future water availability due to climate change, indicate the need to highly prioritize concerted action toward adaptation to climate change in the Mygdonia water system [48–50]. Research on Mygdonia Basin is limited and mainly concerns Lake Koronia, although there are some studies on Lake Volvi that mostly focused on water quality issues [51–56]. Some attempts were made to address simulations the restoration of the water balance of the hydrological basin of Lake Koronia by Manakou [57], while Zalidis [58] and Zalidis et al. [59] studied the Master Plan for the restoration of Lake Koronia. Kolokytha [60–63] examined the impact of WFD and EU CAP and the water footprint of crops in the Lake Koronia basin. Veranis [46] studied the hydrogeology of Mygdonia Basin, while the perspectives of the exploitation of the deep aquifer for the restoration of the Lake Koronia were examined by Mylopoulos et al. [64,65]. Our group has tried an integrated approach that investigated the impacts of climate change in the whole Mygdonia water system (conjunctive use of surface and groundwater resources) and the economy of the region [47,49,66].

2.5. Methodology

- ➢ First, a fully integrated hydrological analysis of the Mygdonia Basin for historical and future periods was carried out. Future climatic data were derived and analyzed from the SMHIRCA Regional Climate Model (chosen among MPI-M-REMO and METO-HC_HardRM3Q0 as the most credible to simulate the P and T of historical data), while climate change impacts on the water balance of both lakes and the Mygdonia Basin aquifer (conjunctive management of surface and groundwater resources) until 2100 were projected by developing a modeling system that included coupled hydrological and hydraulic models, namely, UTHBAL [67,68] MIKE SHE, MIKE HYDRO River and MIKE HYDRO Basin [69–71]. Details can be found in Malamataris et al. [47].

The final outputs of the modeling system (Figure 2) included the future water balance of the Mygdonia Basin aquifer and the Lakes Koronia and Volvi, as well as the piezo-metric surface and the water level, surface area and stored volume of the lakes until 2100. Details can be found in Malamataris et al. [47].

➢ By using the information of the previous hydrological analysis, three sustainable development strategies were formed that combined different adaptation measures seeking sustainable solutions for the restoration of the water system and the development of the area. The results were published in Kolokytha et al. [49]. Some of the results are discussed in this work as an example of adaptation to climate change in water systems with a high negative water balance.

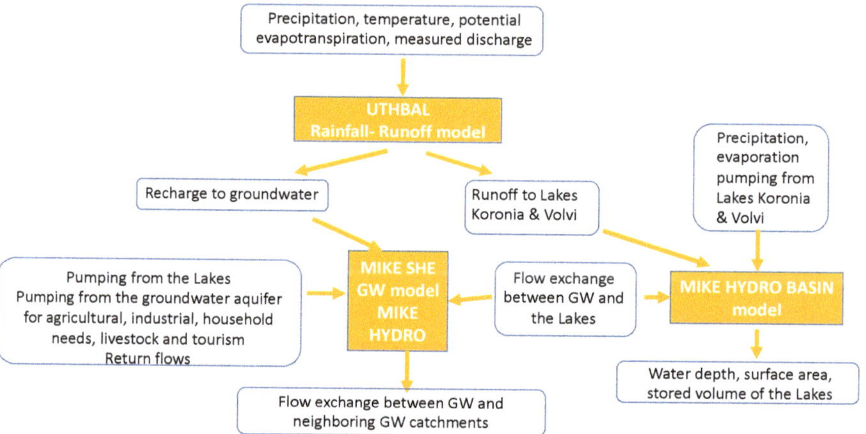

Figure 2. The developed integrated modeling system applied in the Mygdonia Basin that was modified, adopted by Malamataris D. et al. [47]. GW refers to groundwater.

3. Results

In the last few decades, the water balance of Mygdonia Basin has been constantly negative as a result of the fact that water consumption has been constantly exceeding the natural recharge in the basin. From the study of the water balance of the Mygdonia water system, taking into account the historical reference data (1970–2000) and climate projections, it is evident that the water balance is getting worse as the rainfall is anticipated to further decrease (between −5.45% (period 2020–2050) to −17.99% (2050–2080)). Tables 1–3 summarize and reveal the extent of the water problem. The year 2010 was selected to be the starting year for the future climate and hydraulic study because of the lack of measured meteorological data for the period 2000–2010. The small improvement in the groundwater balance during the 2010–2040 period was mainly credited to the significant reduction in the groundwater withdrawals for industrial use due to the shutdown of industries. The vast majority of the industries that were under operation during the historical reference period are currently inactive, mainly because of the pumped water limitations that were specifically established for the Mygdonia catchment. The total amount of pumped water for industrial use is estimated to be reduced by 90% compared with the historical period.

Table 1. Mean water balance of Mygdonia aquifers (hm^3/year).

		1970–2000	2010–2040	2040–2070	2070–2100
Inflows to Mygdonia aquifers (hm^3/year)	Recharge	129.33	127.58	87.95	80.67
	Return flow of the pumped water for irrigation use	24.85	25.49	28.84	30.87
	Inflow from Lake Koronia	0.00	0.00	0.00	0.00
	Inflow from Lake Volvi	0.66	2.25	2.76	3.72
	Inflow from the neighboring groundwater aquifer	0.00	0.00	0.35	2.01
Outflows from Mygdonia aquifers (hm^3/year)	Outflow for agricultural use	164.94	167.51	189.85	203.39
	Outflow for industrial use	7.33	0.73	0.73	0.73
	Outflow for livestock use	1.86	1.86	1.86	1.86
	Outflow for household use	6.40	6.40	6.40	6.40
	Outflow for tourist use	0.02	0.02	0.02	0.02
	Outflow to Lake Koronia	0.00	0.00	0.00	0.00
	Outflow to Lake Volvi	0.41	0.03	0.03	0.00
	Outflow to the neighboring groundwater aquifer	1.35	1.31	0.28	0.00
Mean water balance of Mygdonia aquifers (hm^3/year)		−27.47	−22.54	−79.27	−95.13

Table 2. Mean water balance of Lake Koronia (hm^3/year).

		1970–2000	2010–2040	2040–2070	2070–2100
Inflows to the lake Koronia (hm^3/year)	Precipitation	20.61	19.74	17.05	16.29
	Runoff from the watershed	10.90	15.50	12.13	11.33
	Inflow from the groundwater aquifer	0.00	0.00	0.00	0.00
Outflows from Lake Koronia (hm^3/year)	Evaporation	38.24	33.86	29.49	28.09
	Pumping from the lake	0.00	0.00	0.00	0.00
	Outflow to the groundwater aquifer	0.00	0.00	0.00	0.00
	Overflow to the Derveni stream	0.22	0.00	0.00	0.00
	Mean water balance of Lake Koronia (hm^3/year)	−6.95	+1.38	−0.31	−0.47

In the Mygdonia Basin, different supply- and demand-oriented measures were tested. The supply-oriented measures focused on stream diversion and reservoir construction to enhance the water potential of the water system, which was found to be inadequate to restore the degraded water system. Details can be found in [47].

Three Sustainable Development Strategies (SDS) were formed via the combination of some demand management measures to reduce water use and improve water efficiency through the restructuring of crops and changing irrigation systems based on different local, regional, national and international policies.

In the Mygdonia Basin, the dominant crop is cereal, which is a rain-fed crop, and thus does not negatively affect the water balance of the basin. Maize and alfalfa are the highest water-intensive crops and provide high farm income. Low-water-intensive crops, such as cereals and animal feed, provide low farm income. The area per type of crop from data by the Regional Administration of Central Macedonia in the Mygdonia Basin is depicted in Figure 3. From the net agricultural income per crop from data of the Greek Ministry of Environment, Energy and Climate Change (POL1077/2014), the current annual total net farm income in the whole Mygdonia Basin is estimated to be equal to EUR 45,773,037.90.

Table 3. Mean water balance of Lake Volvi (hm^3/year).

		1970–2000	2010–2040	2040–2070	2070–2100
Inflows to Lake Volvi (hm^3/year)	Precipitation	31.61	31.99	27.27	25.94
	Runoff from the watershed	20.92	15.82	14.65	13.79
	Inflow from the groundwater aquifer	0.41	0.03	0.03	0.00
	Inflow from the Derveni stream	0.22	0.00	0.00	0.00
Outflows from Lake Volvi (hm^3/year)	Evaporation	54.69	48.57	38.72	34.99
	Pumping from the lake	0.70	2.40	2.40	2.40
	Outflow to the groundwater aquifer	0.66	2.25	2.76	3.72
	Outflow to the Richios stream	0.00	0.13	0.00	0.00
Mean water balance of lake Volvi (hm^3/year)		−2.89	−5.51	−1.93	−1.38

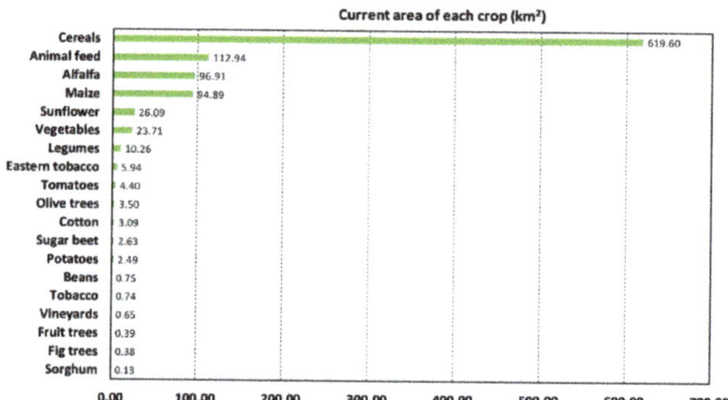

Figure 3. The area of each crop in the Mygdonia Basin.

The three Sustainable Development Strategies tested are:

1. **The reduction of the total irrigated agricultural area (SDS1)** related to a national priority (NP), the Rural Development Plan 2014–2020 of Greece and the EU Common Agricultural Policy, according to which, the fallow land measure was proposed to be applied to at least 30% of the irrigated area, while in the set-aside land, all types of agricultural activities are prohibited. In particular, the fallow land measure was proposed to be applied to all kinds of irrigated crops in the study area with the exception of tree plantations and vineyards, which are multiannual crops. Following a trial-and-error method, it was found that the minimum percentage of the irrigated land of the Mygdonia Basin that should be set aside in order to achieve a water surplus of Mygdonia aquifers is equal to 33% and 61% for the short-term (2020–2050) and long-term (2050–2080) future periods, respectively. By applying the proposed SDS1, the irrigation water demand was projected to decrease from 178.38 hm^3/year to 119.50 hm^3/year in the 2020–2050 period and from 192.91 hm^3/year to 75.38 hm^3/year

in the 2050–2080 period. If SDS1 is applied, the farmers' incomes will have a reduction from EUR 45,773,037.90 to EUR 35,395,528.60 in the short term (2020–2050) and EUR 26,587,206.00 in the long term (2050–2080). More details on crops can be found in Kolokytha et al. [47].

2. **The expansion of the livestock sector as alternative economic activity along with the restructuring of crops (SDS2).**

   Hay production from alfalfa and livestock crops for the needs of the livestock sector is far more (280,201.00 ton/year) than what is needed for the animals (50,327.51 ton/year); therefore, there is significant space to enlarge the livestock sector. SDS2 refers to an increase of 20% of the livestock sector in the Mygdonia Basin based on the report of the Management Body of Lakes Koronia and Volvi. The livestock feed cultivated area, i.e., the alfalfa and the animal feed crops, was proposed to decrease in order to reduce the irrigation water needs in the study area. The maximum acceptable decrease in the livestock feed area was found to be equal to 75%. In the 2020–2050 period, the minimum decrease in the livestock feed area was estimated to be equal to 62% for a water surplus of the Mygdonia aquifers to be achieved. In the 2050–2080 period, even the decrease in the livestock feed area at 75% could not rehabilitate the water deficit of the aquifers; therefore, it was proposed to be accompanied by a reduction in the remaining water-intensive crops in the study area, i.e., maize, tobacco, eastern type tobacco, cotton, sugar beet, sorghum, beans, legumes and potatoes. Of course, the reduction in the irrigated area was projected to decrease the total farm income in the Mygdonia Basin. To compensate for the reduction in rural income, the cut of the irrigated area was proposed to be implemented along with a promotion of crops providing a high net income for farmers so that the economic component of sustainability was also met. The plantation of energy crops for biofuels production, which are projected to be greatly competitive in the future according to European adaptation policies, was tested for the restructuring of crops. Among nine energy crops that were examined in terms of the suitable climate and soil conditions, the irrigation requirements, the harvest period and the net income provided to farmers, *cardoon* is the most suitable one regarding the climatic and environmental conditions of the study area. This particular crop is highly resistant to the Mediterranean climate (high durability in the low temperatures of the Mediterranean winter, high adaptability to arid conditions with prolonged periods of drought) and provides a high net farm income, about 76,533.33 EUR/km^2/year. In particular, cardoon and cereals (rain-fed crops) that improve the food security of the study area were proposed to be promoted.

3. **A combination of the promotion of drip irrigation systems, along with the restructuring of crops (SDS3).** SDS3 concerns the promotion of drip irrigation systems in maize crop cultivation. The future water deficit of the Mygdonia aquifers could not be rehabilitated, even if this measure would be applied throughout the whole Mygdonia Basin. Therefore, the area of the water-intensive crops mentioned in SDS2 needs to be reduced. In the 2020–2050 and 2050–2080 periods, the minimum required decrease in the area of these crops was found to be equal to 35% and 66%, respectively, for the restoration of the Mygdonia aquifers to be achieved. The drip irrigation systems were proposed to be installed at 25% and 35% of the maize cultivated area during the 2020–2050 and 2050–2080 periods, respectively. For the compensation of the farm income loss, crops that provide high net farm income, such as vegetables, potatoes and cardoon, were proposed to be promoted, along with cereals.

The future water balances of the Mygdonia aquifers and Lakes Koronia and Volvi under the Sustainable Development Strategies are presented in Tables 4–9.

Table 4. Average annual water balance of the Mygdonia aquifers for the 2020–2050 period under the SDS1, SDS2, SDS3.

2020–2050 Period	No-Action Scenario	SDS1	SDS2	SDS3
Inflows to Mygdonia aquifers (hm^3/year)				
Recharge	112.68	112.68	112.68	112.68
Return flow of the pumped water for irrigation use	27.12	18.29	18.19	18.07
Inflow from Lake Koronia	0.00	0.00	0.00	0.00
Inflow from Lake Volvi	1.75	0.71	0.75	0.73
Inflow from the neighbouring groundwater aquifer	0.00	0.00	0.00	0.00
Outflows from Mygdonia aquifers (hm^3/year)				
Outflow for agricultural use	178.38	119.50	118.86	118.06
Outflow for industrial use	0.73	0.73	0.73	0.73
Outflow for livestock use	1.86	1.86	2.07	1.86
Outflow for household use	6.40	6.40	6.40	6.40
Outflow for tourist use	0.02	0.02	0.02	0.02
Outflow to Lake Koronia	0.00	0.00	0.00	0.00
Outflow to Lake Volvi	0.06	0.52	0.48	0.49
Outflow to the neighbouring groundwater aquifer	1.15	1.80	1.75	1.74
Water balance of the Mygdonia aquifers (hm^3/year)	−47.05	+0.85	+1.31	+2.18

Table 5. Average annual water balance of the Mygdonia aquifers for the 2050–2080 period under the SDS1–3.

2050–2080 Period	No-Action Scenario	SDS1	SDS2	SDS3
Inflows to Mygdonia aquifers (hm^3/year)				
Recharge	75.50	75.50	75.50	75.50
Return flow of the pumped water for irrigation use	29.30	11.67	11.58	11.49
Inflow from Lake Koronia	0.00	0.00	0.00	0.00
Inflow from Lake Volvi	2.76	0.32	0.35	0.34
Inflow from the neighbouring groundwater aquifer	0.80	0.00	0.00	0.00
Outflows from Mygdonia aquifers (hm^3/year)				
Outflow for agricultural use	192.91	75.38	74.82	74.20
Outflow for industrial use	0.73	0.73	0.73	0.73
Outflow for livestock use	1.86	1.86	2.07	1.86
Outflow for household use	6.40	6.40	6.40	6.40
Outflow for tourist use	0.02	0.02	0.02	0.02
Outflow to Lake Koronia	0.00	0.00	0.00	0.00
Outflow to Lake Volvi	0.00	1.16	1.06	1.05
Outflow to the neighbouring groundwater aquifer	0.02	1.74	1.61	1.59
Water balance of the Mygdonia aquifers (hm^3/year)	−93.58	+0.20	+0.72	+1.48

Table 6. Average annual water balance of the Lake Koronia for the 2020–2050 period under the SDS1–3.

2020–2050 Period	No-Action Scenario	SDS1	SDS2	SDS3
	Inflows to Lake Koronia (hm^3/year)			
Direct precipitation	18.76	18.76	18.76	18.76
Runoff from the watershed	14.48	14.48	14.48	14.48
Inflow from the groundwater aquifer	0.00	0.00	0.00	0.00
	Outflows from Lake Koronia (hm^3/year)			
Evaporation	34.21	34.21	34.21	34.21
Pumping from the lake	0.00	0.00	0.00	0.00
Outflow to the groundwater aquifer	0.00	0.00	0.00	0.00
Overflow to the Derveni stream	0.00	0.00	0.00	0.00
Water balance of the Lake Koronia (hm^3/year)	−0.97	−0.97	−0.97	−0.97

Table 7. Average annual water balance of the Lake Koronia for the 2050–2080 period under the SDS1–3.

2050–2080 Period	No-Action Scenario	SDS1	SDS2	SDS3
	Inflows to Lake Koronia (hm^3/year)			
Direct precipitation	15.89	15.89	15.89	15.89
Runoff from the watershed	10.60	10.60	10.60	10.60
Inflow from the groundwater aquifer	0.00	0.00	0.00	0.00
	Outflows from Lake Koronia (hm^3/year)			
Evaporation	26.81	26.81	26.81	26.81
Pumping from the lake	0.00	0.00	0.00	0.00
Outflow to the groundwater aquifer	0.00	0.00	0.00	0.00
Overflow to the Derveni stream	0.00	0.00	0.00	0.00
Water balance of the Lake Koronia (hm^3/year)	−0.32	−0.32	−0.32	−0.32

Table 8. Average annual water balance of the Lake Volvi for the 2020–2050 period under the SDS1–3.

2020–2050 Period	No-Action Scenario	SDS1	SDS2	SDS3
	Inflows to Lake Volvi (hm^3/year)			
Direct precipitation	30.50	30.50	30.50	30.50
Runoff from the watershed	15.62	15.62	15.62	15.62
Inflow from the groundwater aquifer	0.06	0.52	0.48	0.49
Inflow from the Derveni stream	0.00	0.00	0.00	0.00
	Outflows from Lake Volvi (hm^3/year)			
Evaporation	45.70	46.68	46.62	46.65
Pumping from the lake	2.40	2.40	2.40	2.40
Outflow to the groundwater aquifer	1.75	0.71	0.75	0.73
Overflow to the Richios stream	0.00	0.00	0.00	0.00
Water balance of the Lake Volvi (hm^3/year)	–3.67	–3.15	−3.17	−3.17

Table 9. Average annual water balance of the Lake Volvi for the 2050–2080 period under the SDS1–3.

2050–2080 Period	No-Action Scenario	SDS1	SDS2	SDS3
Inflows to Lake Volvi (hm^3/year)				
Direct precipitation	25.58	25.58	25.58	25.58
Runoff from the watershed	12.94	12.94	12.94	12.94
Inflow from the groundwater aquifer	0.00	1.16	1.06	1.05
Inflow from the Derveni stream	0.00	0.00	0.00	0.00
Outflows from Lake Volvi (hm^3/year)				
Evaporation	36.19	39.59	39.47	39.48
Pumping from the lake	2.40	2.40	2.40	2.40
Outflow to the groundwater aquifer	2.76	0.32	0.35	0.34
Overflow to the Richios stream	0.00	0.00	0.00	0.00
Water balance of the Lake Volvi (hm^3/year)	−2.83	−2.63	−2.64	−2.65

Analysis of the Results

All three Sustainable Development Strategies are projected to achieve a water surplus of the Mygdonia Basin for both the short-term and long-term future periods. Tables 4 and 5 shows that the significant reduction of discharge for agricultural use is the main reason for the quantitative rehabilitation of the basin. The restructuring of the agricultural sector is proposed to be more intense in the 2050–2080 period compared with the 2020–2050 period in order to compensate for the larger decrease in the recharge to aquifers during the long-term future period. In particular, in the 2020–2050 period, the outflow from Mygdonia aquifers for agricultural use is projected to decrease from 178.38 hm^3/year in the no-action scenario to 119.50 hm^3/year (−33.00%) in SDS1, 118.86 hm^3/year (−33.37%) in SDS2 and 118.06 hm^3/year (−33.82%) in SDS3. Furthermore, in the 2050–2080 period, the water withdrawals from the Mygdonia aquifers for irrigation are projected to decrease from 192.91 hm^3/year in the no-action scenario to 75.38 hm^3/year (−60.92%) in SDS1, 74.82 hm^3/year (−61.22%) in SDS2 and 74.20 hm^3/year (−61.54%) in SDS3. The improvement of the water availability of the Mygdonia aquifers is not projected to have a positive effect on the water balance of Lake Koronia (Tables 6 and 7 due to the impermeable bottom of Lake Koronia [68], preventing any water flow exchange between the lake and the aquifers. Moreover, the quantitative rehabilitation of the Mygdonia aquifers is projected to increase the water inflow from the Mygdonia aquifers to Lake Volvi and to decrease the water outflow from the lake to aquifers, resulting in an increase in evaporative losses due to the increase of the water availability and surface area of the lake and, finally, to the slight improvement of the water balance of Lake Volvi, as shown in Tables 8 and 9.

In terms of economic sustainability (rural income), the implementation of the fallow land scheme in SDS1 is expected to cause a decrease in the total net farm income in the Mygdonia Basin from 45,773,037.90 EUR/year (no-action scenario) to 35,395,528.60 EUR/year in 2020–2050 and 26,587,206.00 EUR/year in 2050–2080. A compensatory policy in the form of offsets may be a good option.

Increasing livestock farming while reducing alfalfa farming (water-intensive cultivation) and promoting energy crops proposed in SDS2 provide a competitive alternative growth option that is expected to increase the total net farm income from 45,773,037.90 EUR/year (no-action scenario) to 46,194,872.34 EUR/year (+0.92%) in 2020–2050 and 45,901,602.53 EUR/year (+0.28%) in 2050–2080. This strategy is capable of successfully managing the groundwater aquifer deficit. The promotion of drip irrigation systems, along with the crops restructuring proposed in SDS3, is expected to provide a small increase to the farmers' net income from 45,773,037.90 EUR/year (no-action scenario) to 46,267,432.24 EUR/year (+1.08%) in 2020–2050 and 45,971,157.73 EUR/year (+0.43%) in 2050–2080.

Another adaptation scenario based on the European Energy Roadmap 2050 concerned the exploitation of the wind potential, with the installation of wind farms to produce "green" electric energy that can be used in pumping wells for irrigation. The land-use feasibility and economic viability of this plan were examined using the "Greek Special Framework for Spatial Planning for Renewable Energy Sources", which is the main legislative instrument for establishing wind farms according to land-use criteria. It provides criteria and guidelines for the site allocation of RES projects with an emphasis on wind systems and ensures the sustainability of RES investments through their harmonious incorporation within the natural and human environment. The combination of 10 environmental, 1 technical, 5 economic and 3 social criteria resulted in the determination of an eligible area for establishing wind farms based on land-use criteria. The future wind potential of the Mygdonia Basin was estimated using the monthly wind speed projections at a height of 10 m above the surface. The future wind speed data were derived from the SMHIRCA climate model under the IPCC SRES A1B with a spatial resolution of almost 25 km. The annual wind speed in the whole basin at a 10 m height above the surface was found to be less than 4 m/s, which is insufficient since this wind speed is lower than the cut-in speed of wind turbines. The installation of wind turbines at heights of 80, 100, 120 and 140 m above the surface was examined. Finally, it was found that wind turbines should be established at a height of 140 m above the surface, where the generated electric energy by wind was found to be a non-economically viable plan since the total cost of the installation of the wind turbines was estimated to be far more expensive than that from conventional resources to be used in pumping wells for irrigation. Details on the methodology can be found in [66].

4. Discussion

Mygdonia Basin is a typical Mediterranean agricultural highly negative basin in terms of water balance. This study aimed to summarize and assess potential adaptation measures to reverse the unsustainable water management as a combination of different strategies, which were formed by taking into account both the hydrology of the basin; socioeconomic conditions; environmental parameters; and several international, national, regional and local policies. Sustainable strategies for the restoration of the water bodies in Mygdonia Basin for the 2020–2050 and 2050–2080 periods were tested as adaptation solutions. In addition, the economic effects on the farmers' incomes were estimated for each of the proposed strategies.

It was confirmed that in cases of drained water basins, such as Mygdonia Basin, the efficacy of measures that will only reduce/limit the use of water does not provide a sustainable solution to the problem.

Of all the combined strategies, the most effective one that could guarantee the restoration of both the aquifer and the lakes and assure the sustainability of the water system is the drastic suspension of the agricultural sector. Concurrently, the expansion of alternative competitive economic activities that are capable of maintaining the income of the inhabitants and ensuring the sustainability and protection of water resources was proposed. More specifically, the implementation of a fallow land measure, as well as the promotion of rain-fed or low-water-intensive crops, in combination with the promotion of drip irrigation systems, was found to be the ideal solution for sustainable water management and development in the area under study.

Highly water-stressed hydrological basins, such as Mygdonia Basin, which are also found in other Mediterranean countries, can only be restored if there are drastic reductions in economic water-consuming activities, such as agriculture, or a shift to other new economic activities that are compatible with the new climate conditions.

Generally, the new ominous climate context calls for a new approach. The development of a new "social calculus" that would enable the water community to not only meet present crises but also consider the meaning of our actions in the context of long-range plans could be feasible and sustainable in the future. The change in perception of how water should be treated in all uses is the key to sustainable water management. In the case of farmers, irrigation efficiency is most of the time synonymous with maximizing net revenue rather than saving water. Even in cases of limited supplies, water-saving is not a priority for most farmers. Although innovative irrigation technologies are at the forefront of the solution in saving water in agriculture, labor and other inputs to get better economic gains are the main concern of farmers who make irrigation decisions by relying mostly on their practical experience. The selection of crops used to be related to subsidies coming from the EU, which resulted in the depletion of water resources in many basins with water problems, as water-intensive consuming crops were chosen (cotton, rice, etc.). In such cases, the development of local irrigation advisory services, education and training programs are important and should be considered by the relevant authorities. Farmers generally lack adequate assistance to develop and adopt better approaches and need better technical advice.

Adaptation goes beyond traditional coordination between agencies, interactions between water uses and planning approaches that consider all possible strategies and impacts. It is an integral part of a region's social and economic development. It refers to a norm-shifting change in management and policy that is mainly focused on demand management although strategies for adaptation that entail both water supply and water demand management.

The following directions are central for effective adaptation to climate change in water management.

Change the "Orientation" of the Hydraulic Engineering Projects

Until recently, large infrastructure (dams and reservoirs) were designed and constructed to store excess seasonal water for future use under the condition that the availability of water resources is almost constant over time. Now that we are experiencing a deep change in the pattern of the hydrological cycle with the more frequent occurrence of extreme events and a severe decrease in water reserves, this option can no longer satisfy water demand and competing needs. Adaptive management with regard to hydraulic engineering projects means managing the consequences of extreme events, i.e., the frequent interchange of the period of droughts and floods. More specifically, groundwater recharge management and managed aquifer recharge (MAR) [72–74] are among the most effective engineering techniques. The constant decrease in precipitation and runoff requires a new orientation of water storage engineering projects that will manage the new water deficits. These techniques include the use of treated wastewater for aquifer storage [73–76]. Furthermore, changes in the hydrology of a basin in relation to changes in social values may result in new uses for reservoir storage, which may have greater economic or social value. Of course, high uncertainty in future projections for water availability poses a threat to the accuracy of projections.

Moreover, the adoption of an integrated approach is required to manage the consequences of natural disasters (fires, droughts and floods) in terms of providing proactive measures to prevent disasters and shield natural systems, such as through fire protection, control of forest fires, erosion prevention and flood control projects.

Given that interventions in existing infrastructure to enhance water resources have limited potential, due to climate projections that predict a decrease in rainfall, the only effective and sustainable way out is to adopt demand management methods and measures. Demand management measures consist of a great array of techniques and tools (engineering, economic, environmental, institutional and social). Demand management may be applied through the implementation of strict standards in terms of water use and appropriate legislation to limit and control illegal drilling. In Mediterranean countries, such as France, Portugal, Italy and Spain, they have implemented different tax systems on agricultural water abstractions to recover the costs of the regulation, storage and management of basin-level water services with various levels of cost recovery in accordance with the provision in the Water Framework Directive. Meanwhile, incentives, such as tax reduction, would encourage water saving in all uses. Technical interventions to reduce losses in irrigation networks and water efficiency advantages through education and training campaigns on farms are among the options. Economic instruments, such as water metering and water pricing, are critical but very difficult to be implemented, first, because there is a high initial cost for metering that needs to be subsided by the authorities and, second, because the appropriate pricing policy needs detailed analysis that takes into account socioeconomic constraints, among others, which, for the moment in Greece, is still in the early stage. Educating water users, giving access to data and strengthening the cooperation between academia and industry are also important elements for successful demand management.

All the measures are location- and case-specific since adaptation is more effective at the local level. In the areas most vulnerable to climate change, emphasis is given to socio-economic development and robust agricultural management. Sometimes it is wise to make a combination of adopting traditional techniques of water storage (cisterns) and traditional cultivations, which are already well known to farmers, with the use of modern technology to achieve water conservation. Moreover, the choice of crops and crop pattern in the case of irrigated agriculture should be compatible with the climatic conditions of the area and the land (soil). More efficient irrigation systems, shifts toward less water-intensive or more drought-tolerant crops, application of deficit irrigation schemes, land reclamation and land management for carbon sequestration [77–80] may reduce water needs for agriculture and increase water-use efficiency [27,75]. The yearly water withdrawal for irrigation in the Mediterranean region amounts to ~223 km^3 [81], but there is a great water saving potential through the implementation of efficient irrigation systems [82–84]. The rebound effect should be mentioned though, as the water-saving effect of efficient irrigation systems may be counterbalanced by the expansion of irrigated areas [63,85,86].

5. Conclusions

Generally, highly water-stressed hydrological basins, such as Mygdonia Basin, which are also found in other Mediterranean countries, can only be restored if there are drastic reductions in water-consuming economic activities (agriculture) and a change in the development model in such a way to shift to other alternative development plans that are compatible with the new climate conditions.

Although great emphasis must be placed on water management and irrigation efficiency, which is still not done in the best way today (many parts of the country are already facing serious problems), this option alone cannot alleviate the water problems.

The reformation of the development plans in the agricultural sector, which is a global issue, is tightly connected with food security, which should be a priority in every agricultural basin. To assure food security, a set of crops that are compatible with the climate, soil and water requirements of the area under study should be prioritized.

All these interventions/changes should be accompanied by information campaigns for growers and breeders on both the effects of climate change on their work and the tools available to manage these effects. Training is important for implementing new crop diversification and rotation, the selection of crops that are better adapted to the new climatic conditions for the region in question and the targeted drainage of agricultural land.

Above all, political will and governmental engagement are crucial for the adoption of a new development model based on the available water resources and the comparative advantages of each area. In the case of highly water-stressed basins, agriculture should not be a priority. Seeking alternative economic activities that use less water is the best adaptation.

The methodology followed in this research, as well as the proposed sustainable development strategies, could prove very useful, not only for the local authorities in Mygdonia Basin but also for any other river basin with similar characteristics.

Funding: This research received no external funding.

Institutional Review Board Statement: Not applicable.

Informed Consent Statement: Not applicable.

Data Availability Statement: This is an article. Some of the results were published in Malamataris, D.; Kolokytha, E.; Loukas, A. Integrated hydrological modeling of surface water and groundwater under climate change: The case of Mygdonia Basin in Greece, *J. Water Clim. Change* **2020**, *11*, 1429–1454. https://doi.org/10.2166/wcc.2019.011 (accessed on 27 December 2021), and Kolokytha, E.; Malamataris, D. Integrated Water Management Approach for Adaptation to Climate Change in Highly Water Stressed Basins, *Water Resour. Manage.* **2020**, *34*, 1173–1197, https://doi.org/10.1007/s11269-020-02492-w (accessed on 27 December 2021).

Acknowledgments: The results of this research are part of a doctoral thesis (Malamataris 2019), "Integrated water resources management and sustainable development strategy in a changing climate: Case study of the Mygdonia water basin" which was funded by the IKY Fellowships of Excellence for Postgraduate Studies in Greece—Siemens Program. Special thanks to DHI for the kind provision of the software MIKE SHE, MIKE HYDRO River and MIKE HYDRO Basin licenses that were used for the modeling tasks that are mentioned in this paper.

Conflicts of Interest: The author declares no conflict of interest.

References

1. Koutsoyiannis, D. Hurst-Kolmogorov Dynamics and Uncertainty. *JAWRA J. Am. Water Resour. Assoc.* **2011**, *47*, 481–495. [CrossRef]
2. Cook, J.; Oreskes, N.; Doran, P.T.; Anderegg, W.R.L.; Verheggen, B.; Maibach, E.W.; Carlton, J.S.; Lewandowsky, S.; Skuce, A.G.; Green, S.A.; et al. Consensus on Consensus: A Synthesis of Consensus Estimates on Human-Caused Global Warming. *Environ. Res. Lett.* **2016**, *11*, 048002. [CrossRef]
3. Lins, H.F.; Cohn, T.A. Stationarity: Wanted Dead or Alive? *JAWRA J. Am. Water Resour. Assoc.* **2011**, *47*, 475–480. [CrossRef]
4. Masson-Delmotte, V.; Zhai, P.; Chen, Y.; Goldfarb, L.; Gomis, M.I.; Matthews, J.B.R.; Berger, S.; Huang, M.; Yelekçi, O.; Yu, R.; et al. *Working Group I Contribution to the Sixth Assessment Report of the Intergovernmental Panel on Climate Change*; Cambridge University Press: Cambridge, UK, 2021.
5. UN Office. *UNDRR Global Natural Disaster Assessment Report*; UN Annual Report; UN Office: Geneva, Switzerland, 2020.
6. Climate Change | UN-Water. Available online: https://www.unwater.org/water-facts/climate-change/ (accessed on 27 December 2021).
7. Fawzy, S.; Osman, A.I.; Doran, J.; Rooney, D.W. Strategies for Mitigation of Climate Change: A Review. *Environ. Chem. Lett.* **2020**, *18*, 2069–2094. [CrossRef]
8. Noble, I.R.; Huq, S.; Anokhin, Y.A.; Carmin, J.A.; Goudou, D.; Lansigan, F.P.; Osman-Elasha, B.; Villamizar, A.; Patt, A.; Takeuchi, K.; et al. Adaptation Needs and Options. In *Climate Change 2014: Impacts, Adaptation, and Vulnerability. Part A: Global and Sectoral Aspects*; Cambridge University Press: Cambridge, UK, 2014; pp. 833–868.
9. Olmstead, S.M. Climate Change Adaptation and Water Resource Management: A Review of the Literature. *Energy Econ.* **2014**, *46*, 500–509. [CrossRef]
10. Bours, D.; Mcginn, C.; Pringle, P. *Monitoring & Evaluation for Climate Change Adaptation: A Synthesis of Tools, Frameworks and Approaches*; SEA Change, CoP, Phnom Penh and UKCIP: Oxford, UK, 2013.
11. Smit, B.; Burton, I.; Klein, R.J.T.; Wandel, J. An Anatomy of Adaptation to Climate Change and Variability. *Clim. Change* **2000**, *45*, 223–251. [CrossRef]
12. Hallegatte, S. Strategies to Adapt to an Uncertain Climate Change. *Glob. Environ. Change* **2009**, *19*, 240–247. [CrossRef]
13. Adger, W.N.; Lorenzoni, I.; O'Brien, K.L. *Adapting to Climate Change: Thresholds, Values, Governance*; Cambridge University Press: Cambridge, UK, 2009.
14. Salerno, F. Adaptation Strategies for Water Resources: Criteria for Research. *Water* **2017**, *9*, 805. [CrossRef]
15. Adger, W.N.; Arnell, N.W.; Tompkins, E.L. Successful Adaptation to Climate Change across Scales. *Glob. Environ. Change* **2005**, *15*, 77–86. [CrossRef]
16. Thompson, S.E.; Sivapalan, M.; Harman, C.J.; Srinivasan, V.; Hipsey, M.R.; Reed, P.; Montanari, A.; Blöschl, G. Developing Predictive Insight into Changing Water Systems: Use-Inspired Hydrologic Science for the Anthropocene. *Hydrol. Earth Syst. Sci.* **2013**, *17*, 5013–5039. [CrossRef]
17. Pahl-Wostl, C. Transitions towards Adaptive Management of Water Facing Climate and Global Change. *Water Resour. Manag.* **2007**, *21*, 49–62. [CrossRef]
18. Medema, W.; Mcintosh, B.S.; Jeffrey, P.J. From Premise to Practice: A Critical Assessment of Integrated Water Resources Management and Adaptive Management Approaches in the Water Sector. *Ecol. Soc.* **2008**, *13*, 18. [CrossRef]
19. Wilby, R.L.; Dessai, S. Robust Adaptation to Climate Change. *Weather* **2010**, *65*, 180–185. [CrossRef]
20. Kim, Y.O.; Chung, E.S. Adaptation to Climate Change: Decision Making. *Sustain. Water Resour. Plan. Manag. Under Clim. Change* **2016**, *8*, 189–221. [CrossRef]
21. AQUASTAT Database Database Query Results. Available online: https://www.fao.org/aquastat/statistics/query/results.html (accessed on 7 December 2021).
22. Sofios, S.; Arabatzis, G.; Baltas, E. Policy for Management of Water Resources in Greece. *Environmentalist* **2008**, *28*, 185–194. [CrossRef]
23. Mimikou, M.A. Water Resourses in Greece: Present and Future. *Glob. NEST J.* **2005**, *7*, 313–322.
24. Mylopoulos, Y.; Kolokytha, E.; Tolikas, D. Urban Water Management in Greece. *Water Int.* **2003**, *28*, 43–51. [CrossRef]

25. Implementation of River Basin Management Plans—Greece—Environment—European Commission. Available online: https://ec.europa.eu/environment/water/participation/map_mc/countries/greece_en.htm (accessed on 8 December 2021).
26. Kondolf, G.M.; Podolak, K.; Grantham, T.E. Restoring Mediterranean-Climate Rivers. *Hydrobiologia* **2013**, *719*, 527–545. [CrossRef]
27. Cramer, W.; Guiot, J.; Fader, M.; Garrabou, J.; Gattuso, J.P.; Iglesias, A.; Lange, M.A.; Lionello, P.; Llasat, M.C.; Paz, S.; et al. Climate Change and Interconnected Risks to Sustainable Development in the Mediterranean. *Nat. Clim. Change* **2018**, *8*, 972–980. [CrossRef]
28. Barkhordarian, A.; von Storch, H.; Bhend, J. The Expectation of Future Precipitation Change over the Mediterranean Region Is Different from What We Observe. *Clim. Dyn.* **2013**, *40*, 225–244. [CrossRef]
29. Giorgi, F.; Lionello, P. Climate Change Projections for the Mediterranean Region. *Glob. Planet. Change* **2008**, *63*, 90–104. [CrossRef]
30. Lionello, P.; Scarascia, L. The Relation between Climate Change in the Mediterranean Region and Global Warming. *Reg. Environ. Change* **2018**, *18*, 1481–1493. [CrossRef]
31. Toreti, A.; Naveau, P.; Zampieri, M.; Schindler, A.; Scoccimarro, E.; Xoplaki, E.; Dijkstra, H.A.; Gualdi, S.; Luterbacher, J.; Naveau, P.; et al. Projections of Global Changes in Precipitation Extremes from Coupled Model Intercomparison Project Phase 5 Models. *Geophys. Res. Lett.* **2013**, *40*, 4887–4892. [CrossRef]
32. Toreti, A.; Naveau, P. On the Evaluation of Climate Model Simulated Precipitation Extremes. *Environ. Res. Lett.* **2015**, *10*, 014012. [CrossRef]
33. Garrote, L.; Iglesias, A.; Granados, A.; Mediero, L.; Martin-Carrasco, F. Quantitative Assessment of Climate Change Vulnerability of Irrigation Demands in Mediterranean Europe. *Water Resour. Manag.* **2015**, *29*, 325–338. [CrossRef]
34. Charles, H.; Godfray, H.; Garnett, T. Food Security and Sustainable Intensification. *Philos. Trans. R. Soc. B Biol. Sci.* **2014**, *369*. [CrossRef]
35. Döll, P.; Jiménez-Cisneros, B.; Oki, T.; Arnell, N.W.; Benito, G.; Cogley, J.G.; Jiang, T.; Kundzewicz, Z.W.; Mwakalila, S.; Nishijima, A. Integrating Risks of Climate Change into Water Management. *Hydrol. Sci. J.* **2014**, *60*, 4–13. [CrossRef]
36. World Resource Institut, RELEASE: Updated Global Water Risk Atlas Reveals Top Water-Stressed Countries and States. 2019. Available online: https://www.wri.org/news/release-updated-global-water-risk-atlas-reveals-top-water-stressed-countries-and-states (accessed on 27 December 2021).
37. Climate Change Impact Scientific Committee (CCISC). *The Environmental, Economic and Social Impacts of Climate Change in Greece*; Bank of Greece: Athens, Greece, 2011.
38. Tsiros, I.X.; Nastos, P.; Proutsos, N.D.; Tsaousidis, A. Variability of the Aridity Index and Related Drought Parameters in Greece Using Climatological Data over the Last Century (1900–1997). *Atmos. Res.* **2020**, *240*, 104914. [CrossRef]
39. Kourgialas, N.N.; Anyfanti, I.; Karatzas, G.P.; Dokou, Z. An Integrated Method for Assessing Drought Prone Areas—Water Efficiency Practices for a Climate Resilient Mediterranean Agriculture. *Sci. Total Environ.* **2018**, *625*, 1290–1300. [CrossRef]
40. Kourgialas, N.N. A Critical Review of Water Resources in Greece: The Key Role of Agricultural Adaptation to Climate-Water Effects. *Sci. Total Environ.* **2021**, *775*, 145857. [CrossRef]
41. Nastos, P.T.; Zerefos, C.S. Spatial and Temporal Variability of Consecutive Dry and Wet Days in Greece. *Atmos. Res.* **2009**, *94*, 616–628. [CrossRef]
42. Nastos, P.T.; Kapsomenakis, J. Regional Climate Model Simulations of Extreme Air Temperature in Greece. Abnormal or Common Records in the Future Climate? *Atmos. Res.* **2015**, *152*, 43–60. [CrossRef]
43. The Impact of Climate Change on the Greek Economy—Dianeosis. Available online: https://www.dianeosis.org/en/2017/08/impact-climate-change-greek-economy/ (accessed on 8 December 2021).
44. Environplan, S.A. *Regional Adaptation Plan to Climate Change in Central Macedonia*; Technical Report for Thessaloniki: Thessaloniki, Greece, 2021.
45. Bobori, D.; Tsavdaroglou, F.; Bafeiadou, A.; Patsia, A.; Karta, E.; Kravva, M.; Tekidis, E.; Koulogiannis, S.; Avramidis, A.; Poulia, A.; et al. *Annual Report of the Management Body of Lakes Koronia and Volvi*; Technical Report for the Management Body of Lakes Koronia and Volvi: Lagkadas, Greece, 2016; p. 35. (In Greek)
46. Veranis, N. *Hydrogeological Study of Mygdonias, Mavroudas, Agiou Orous, Xolomonta, Olympiadas, Asprolakka, Ierissou, and Rodon Groundwater Systems*; Technical Report for Groundwater Systems: Thessaloniki, Greece, 2010. (In Greek)
47. Malamataris, D.; Kolokytha, E.; Loukas, A. Integrated Hydrological Modelling of Surface Water and Groundwater under Climate Change: The Case of the Mygdonia Basin in Greece. *J. Water Clim. Change* **2020**, *11*, 1429–1454. [CrossRef]
48. Greek Ministry of the Environment. *Energy and Climate Change, Analysis of Pressures and Impacts on Water Resources*; Technical Report for Greek Ministry of the Environment: Athens, Greece, 2020; p. 212. (In Greek)
49. Kolokytha, E.; Malamataris, D. Integrated Water Management Approach for Adaptation to Climate Change in Highly Water Stressed Basins. *Water Resour. Manag.* **2020**, *34*, 1173–1197. [CrossRef]
50. Malamataris, D.; Kolokytha, E.; Mylopoulos, I.; Loukas, A. Critical Review of Adaptation Strategies for the Restoration of Lake Koronia. *Eur. Water* **2017**, *58*, 203–208.
51. Gantidis, N.; Pervolarakis, M.; Fytianos, K. Assessment of the Quality Characteristics of Two Lakes (Koronia and Volvi) of N. Greece. *Environ. Monit. Assess.* **2007**, *125*, 175–181. [CrossRef]
52. Perivolioti, T.; Doxani, G.; Bobori, D.; Perivolioti, T.-M.; Mouratidis, A. Monitoring the water quality of lake koronia using long time-series of multispectral satellite images remote sensing (mdpi) journal: Special issue "remote sensing of soil erosion" view project sentinel-2 view project monitoring the water quality of lake koronia using long time-series of multispectral satellite images. *AUC Geogr. Praha Czech Repub.* **2016**, *54124*, 9–13.
53. Zacharias, I.; Bertachas, I.; Skoulikidis, N.; Koussouris, T. Greek Lakes: Limnological Overview. *Lakes Reserv. Res. Manag.* **2002**, *7*, 55–62. [CrossRef]

54. Christophoridis, C.; Fytianos, K. Conditions Affecting the Release of Phosphorus from Surface Lake Sediments. *J. Environ. Qual.* **2006**, *35*, 1181–1192. [CrossRef]
55. Petaloti, C.; Voutsa, D.; Samara, C.; Sofoniou, M.; Stratis, I.; Kouimtzis, T. Nutrient Dynamics in Shallow Lakes of Northern Greece. *Environ. Sci. Pollut. Res.* **2004**, *11*, 11–17. [CrossRef]
56. Moustaka-Gouni, M.; Michaloudi, E.; Kormas, K.; Katsiapi, M.; Moustaka-Gouni, M.; Michaloudi, E.; Kormas, K.A.; Katsiapi, M.; Vardaka, E.; Genitsaris, S. Plankton Changes as Critical Processes for Restoration Plans of Lakes Kastoria and Koronia Dietary Supplements of Spirulina: Apart from Knowing Benefits Are Any Hidden Treats? View Project LAKEREMAKE View Project 76 PUBLICATIONS 1328 CITATIONS SEE PROFILE Plankton Changes as Critical Processes for Restoration Plans of Lakes Kastoria and Koronia. *Eur. Water* **2012**, *40*, 43–51.
57. Manakou, V.; Tsiakis, P.; Kungolos, A. A Mathematical Programming Approach to Restore the Water Balance of the Hydrological Basin of Lake Koronia. *Desalination Water Treat.* **2013**, *51*, 2955–2976. [CrossRef]
58. Zalidis, G. *Environmental Rehabilitation Plan of Lake Koronia and the Greater Area of Mygdonia Basin*; Technical Report for Thessaloniki: Thessaloniki, Greece, 2004. (In Greek)
59. Zalidis, G.; Alexandridis, T. *Revised Restoration Plan of Lake Koronia*; Technical Report for Thessaloniki: Thessaloniki, Greece, 2004. (In Greek)
60. Kolokytha, E. European Policies for Confronting Yhe Challenges of Climate Change in Water Resources. *Curr. Sci.* **2010**, *98*, 1069–1076.
61. Kolokytha, E. Agricultural Development in Lake Koronia. The Role of the Water Footprint of Major Crops in Combating Climate Change. In Proceedings of the 3rd IAHR Europe Congress, Porto, Portugal, 14–16 April 2014.
62. Kolokytha, E.; Ntota, A.; Mylopoulos, I. Impact of Climate Change Un Greece—The Lake Koronia Case. In Proceedings of the 33rd Inrernational Congress IHAR, e-Proceedings, Vancouver, BC, Canada, 9–14 August 2009.
63. Oishi, S.; Teegavarapu, R.S.V. *Sustainable Water Resources Planning and Management under Climate Change*; Springer: Singapore, 2016; ISBN 9789811020513.
64. Mylopoulos, N.; Mylopoulos, Y.; Kolokytha, E.; Tolikas, D. Integrated Water Management Plans for the Restoration of Lake Koronia, Greece. *Water Int.* **2011**, *32*, 720–738. [CrossRef]
65. Mylopoulos, I.; Tolikas, D.; Mylopoulos, N.; Zorba, A.; Palaiologos, V.; Kolokytha, E.; Mentes, A.; Karamanlidou, M. *Study of the Exploitation Potential of the Deep Aquifer of the Lake Koronia sub-Catchment in the Prefecture of Thessaloniki*; Technical Report for Thessaloniki: Thessaloniki, Greece, 2002; p. 183. (In Greek)
66. Malamataris, D.; Kolokytha, E.; Loukas, A.; Mylopoulos, Y. "Green" Energy in the Framework of the Sustainable Development of the Mygdonia Basin, Greece. In Proceedings of the 7th International Conferenceon Environmental Management, Engineering, Planning, and Economics CEMEPE & SECOTOX Conference, Mykonos, Greece, 19–24 May 2019.
67. Loukas, A.; Mylopoulos, N.; Vasiliades, L. A Modeling System for the Evaluation of Water Resources Management Strategies in Thessaly, Greece. *Water Resour. Manag.* **2007**, *21*, 1673–1702. [CrossRef]
68. Sidiropoulos, P.; Mylopoulos, N.; Loukas, A. Optimal Management of an Overexploited Aquifer under Climate Change: The Lake Karla Case. *Water Resour. Manag. Int. J. Publ. Eur. Water Resour. Assoc. (EWRA)* **2013**, *27*, 1635–1649. [CrossRef]
69. Danish Hydraulic Institut (DHI), MIKE SHE Documentation. Available online: https://manuals.mikepoweredbydhi.help/2017/MIKE_SHE.htm (accessed on 7 December 2021).
70. Danish Hydraulic Institut (DHI), MIKE HYDRO Documentation. Available online: https://manuals.mikepoweredbydhi.help/2017/MIKE_Hydro.htm (accessed on 7 December 2021).
71. Danish Hydraulic Institut (DHI), MIKE HYDRO, Basin User Guide. Available online: https://www.mikepoweredbydhi.com/products/mike-hydro-basin (accessed on 7 December 2021).
72. Zhang, H.; Xu, Y.; Kanyerere, T. A Review of the Managed Aquifer Recharge: Historical Development, Current Situation and Perspectives. *Phys. Chem. Earth Parts A/B/C* **2020**, *118–119*, 102887. [CrossRef]
73. Dillon, P.; Pavelic, P.; Page, D.; Beringen, H.; Ward, J. *Managed Aquifer Recharge, an Introduction*; No.13 Series; CSIRO on behalf of National Water Commission: Canberra, Australia, 2009.
74. Dillon, P.; Arshad, M. Managed Aquifer Recharge in Integrated Water Resource Management. In *Integrated Groundwater Management Concepts, Approaches and Challenges*; Springer: Cham, Switzerland, 2016; pp. 435–452. [CrossRef]
75. Djuma, H.; Bruggeman, A.; Eliades, M.; Lange, M.A. Non-Conventional Water Resources Research in Semi-Arid Countries of the Middle East. *Desalination Water Treat.* **2016**, *57*, 2290–2303. [CrossRef]
76. Barnett, S.R.; Howles, S.R.; Martin, R.R.; Gerges, N.Z. Aquifer Storage and Recharge: Innovation in Water Resources Management. *Aust. J. Earth Sci.* **2000**, *47*, 13–19. [CrossRef]
77. Missimer, T.M.; Drewes, J.E.; Amy, G.; Maliva, R.G.; Keller, S. Restoration of Wadi Aquifers by Artificial Recharge with Treated Waste Water. *Ground Water* **2012**, *50*, 514–527. [CrossRef] [PubMed]
78. Ghaffour, N.; Missimer, T.M.; Amy, G.L. Combined Desalination, Water Reuse, and Aquifer Storage and Recovery to Meet Water Supply Demands in the GCC/MENA Region. *Desalination Water Treat.* **2013**, *51*, 38–43. [CrossRef]
79. Nikolaou, G.; Neocleous, D.; Christou, A.; Kitta, E.; Katsoulas, N. Implementing Sustainable Irrigation in Water-Scarce Regions under the Impact of Climate Change. *Agronomy* **2020**, *10*, 1120. [CrossRef]

80. Almagro, M.; de Vente, J.; Boix-Fayos, C.; García-Franco, N.; Melgares de Aguilar, J.; González, D.; Solé-Benet, A.; Martínez-Mena, M. Sustainable Land Management Practices as Providers of Several Ecosystem Services under Rainfed Mediterranean Agroecosystems. *Mitig. Adapt. Strateg. Glob. Change* **2016**, *21*, 1029–1043. [CrossRef]
81. Fader, M.; Giupponi, C.; Burak, S.; Dakhlaoui, H.; Koutroulis, A.; Lange, M.A.; Llasat, M.C.; Pulido-Velazquez, D.; Sanz-Cobeña, A. Water. In *Climate and Environmental Change in the Mediterranean Basin—Current Situation and Risks for the Future. First Mediterranean Assessment Report*; Cramer, W., Guiot, J., Marini, K., Eds.; Union for the Mediterranean, Plan Bleu, UNEP/MAP: Marseille, France, 2020.
82. Fader, M.; Shi, S.; von Bloh, W.; Bondeau, A.; Cramer, W. Mediterranean Irrigation under Climate Change: More Efficient Irrigation Needed to Compensate for Increases in Irrigation Water Requirements. *Hydrol. Earth Syst. Sci.* **2016**, *20*, 953–973. [CrossRef]
83. Harmanny, K.S.; Malek, Ž. Adaptations in Irrigated Agriculture in the Mediterranean Region: An Overview and Spatial Analysis of Implemented Strategies. *Reg. Environ. Change* **2019**, *19*, 1401–1416. [CrossRef]
84. Fishman, R.; Devineni, N.; Raman, S. Can Improved Agricultural Water Use Efficiency Save India's Groundwater? *Environ. Res. Lett.* **2015**, *10*, 084022. [CrossRef]
85. Song, J.; Guo, Y.; Wu, P.; Sun, S.H. The Agricultural Water Rebound Effect in China. *Ecol. Econ.* **2018**, *146*, 497–506. [CrossRef]
86. Paul, C.; Techen, A.K.; Robinson, J.S.; Helming, K. Rebound Effects in Agricultural Land and Soil Management: Review and Analytical Framework. *J. Clean. Prod.* **2019**, *227*, 1054–1067. [CrossRef]

MDPI
St. Alban-Anlage 66
4052 Basel
Switzerland
Tel. +41 61 683 77 34
Fax +41 61 302 89 18
www.mdpi.com

Water Editorial Office
E-mail: water@mdpi.com
www.mdpi.com/journal/water

www.ingramcontent.com/pod-product-compliance
Lightning Source LLC
LaVergne TN
LVHW070708170726
843515LV00004B/517